global perspectives
in environmental
adult education

Studies in the
Postmodern Theory of Education

Joe L. Kincheloe and Shirley R. Steinberg
General Editors

Vol. 230

PETER LANG
New York • Washington, D.C./Baltimore • Bern
Frankfurt am Main • Berlin • Brussels • Vienna • Oxford

global perspectives in environmental adult education

EDITED BY

Darlene E. Clover

WITH THE ASSISTANCE OF

Sandra Tan

PETER LANG
New York • Washington, D.C./Baltimore • Bern
Frankfurt am Main • Berlin • Brussels • Vienna • Oxford

Library of Congress Cataloging-in-Publication Data

Global perspectives in environmental adult education / edited by Darlene E. Clover
p. cm. — (Counterpoints; vol. 230)
Includes bibliographical references and index.
1. Environmental education. 2. Adult education. I. Clover, Darlene Elaine.
II. Counterpoints (New York, N.Y.); v. 230.
GE70 .G563   363.7'071—dc21   2002070488
ISBN 0-8204-6198-9
ISSN 1058-1634

Die Deutsche Bibliothek-CIP-Einheitsaufnahme

Global perspectives in environmental adult education / ed. by: Darlene E. Clover
–New York; Washington, D.C./Baltimore; Bern;
Frankfurt am Main; Berlin; Brussels; Vienna; Oxford: Lang.
([Counterpoints; Vol. 230)
ISBN 0-8204-6198-9

Cover design by Joni Holst

The paper in this book meets the guidelines for permanence and durability
of the Committee on Production Guidelines for Book Longevity
of the Council of Library Resources.

© 2004 Peter Lang Publishing, Inc., New York
275 Seventh Avenue, 28th Floor, New York, NY 10001
www.peterlangusa.com

Printed in the United States of America

Contents

INTRODUCTION

Darlene E. Clover
University of Victoria, Canada

> New patterns of thought and belief are emerging that will transform our experience,
> our thinking and our action. (Reason 1998, 42)

This book has undergone a number of transformations in the course of its development—a process that reflects a longer journey to develop environmental learning strategies, theoretical and philosophical foundations for an emerging stream of adult education.

This collection is rooted in the work of the Learning for Environmental Action Programme (LEAP) of the International Council for Adult Education (ICAE). Created in 1989, LEAP was the only global network that provided a place of encounter for adult educators who worked within an ecological framework, in community, non-governmental, and inter-governmental organizations, unions, health care facilities, governments, universities and colleges. Through seminars, workshops, and publications, LEAP became a space to articulate the theory and practice of environmental adult education. In 1992, during the Rio Earth Summit under the leadership of Moema Viezzer of Brazil and Darlene E. Clover of Canada, the programme organized the International Journey for Environmental Education, and the creative process which resulted in the first 'treaty' on environmental education entitled "Environmental Education for Sustainable Societies and Global Responsibility" (NGO FORUM/LEAP, 1992). In 1997, LEAP, for the first time in history, placed environmental issues on the agenda of The Fifth International Conference on Adult Education (CONFINTEA V). For over ten years LEAP members took part in numerous global conferences on environmental education and demonstrated the value and need for adult education and learning—for creating a fully active citizenship—based on the belief that

> if a water jug is full of holes, it needs every finger to make it watertight. Our world
> is just like such a jug: broken and full of holes. If every one of us does not make the
> necessary contribution, then this world will perish. (Mané 1993, 7)

Although the programme has come to end, the work continues in different forms. This book is a contribution to that ongoing process.

Adult Education and Learning

The nature of adult education is essentially changing as new ideas and purposes are grafted onto existing lines of evolution and development. These general images also support the elastic and wider interpretations of the meaning of adult education. (Elsey 1986, 11)

The field of adult education and learning is complex, dynamic, and everchanging. The term can be used to describe an area of academic inquiry, learning processes within social movements, non-formal learning activities, continuing education and training (Selman and Dampier 1991; Grace 1998). A frequently cited definition of adult education comes from the United Nations Educational, Scientific and Cultural Organization (UNESCO). UNESCO notes that adult education should be understood as:

the entire body of organised educational process, whatever the content, level and method, whether formal or otherwise, whether they prolong or replace initial education in schools, colleges and universities as well as in apprenticeship, whereby persons regarded as adult by the society to which they belong develop their abilities, enrich their knowledge, improve their technical or professional qualifications or turn them in a new direction and bring about changes in their attitudes or behaviour in the twofold prospect of full personal development and participation in balanced independent social, economic and cultural development. (1976, 3)

Over the years, a more critical and social justice-oriented adult education, the stream in which the authors of this book work, has emerged. This vision of adult education is embedded within the process of daily living and has societal transformation through collective learning and action as its goal.

Critical adult education begins with peoples' daily lived experiences, the bedrock upon which meaning is constructed. The essential strategy is to take up social, cultural, historical, economic, and political considerations to radically question "the status quo, corporate capitalism and threats to social democracy" (Grace 1998, 117). Adult education works to help people reflect upon how "ideological systems and societal structures hinder or impede the fullest development of humankind's collective potential" (Welton 1995, 14) through "collective action...community building...dialogue, communication, conflict and change" (Grace 1998, 117).

Feminist adult educators took the discourse one step further with an analysis of oppression from the perspective of women. While agreeing with the view that education and learning should be critical tools for social change, they demonstrated how the universalizing tendencies and political assumptions generated by theories of adult education tended to be "male-dominated and gender insensitive" (Luke and Gore 1992, 8). Grounded in a rationalist, masculinist ideology, they are far from 'liberatory practices' for women (Luke and Gore 1992, ix) but rather inadvertently maintain a "gendered division between male public and female private, culture and nature, reason and emotion, and mind and body" (Luke and Gore 1992, 34). Feminist adult education is an educational practice "that works as a mobilizing axis for social change and consciousness-building around gender" and other forms of oppression (Gina Vargas quoted in Rosero 1993, 80).

Adult education goes well "beyond conventional and narrowly defined educational considerations into those which encompass welfare, employment, race and gender issues" (Bryant and Usher, quoted in Thomas 1991, 302). However, it is only recently that the field has begun to engage in the environment discourse. Environmental adult education offers a new weave to the ecologically threadbare carpet of adult education, recognizing the rest of nature and the politics of the environment as a complex and integral segment of the fabric of human existence on this planet.

The Environmental Challenge

> The argument over what we may and may not teach in order to reconcile socio-economic development with environmental protection and rehabilitation (in short, quality of life), and to accommodate scientific exploration and uncertainty leads to the need to identify the broader context in which we teach. (Sutton 1989, 8)

Over the past three decades, broader attention has been given to analyzing and debating the environmental crisis. Increasingly, much intellectual and creative energy has been placed in defining the environment as an integrated human and non-human context or space within which all thought, emotion and activity must also be situated. There has been a creative retreat from viewing environmental problems as just about the natural world, toward an understanding of its undeniable relevance within the mosaic of the social, political, economic, cultural and personal landscapes of life. While historians re-write history through an ecological lens, sociologists and anthropologists develop new ecological frameworks for social and cultural activity. While women worldwide engage in direct action and conceptualize theories of eco-feminism, aboriginal people teach traditional ecological knowledge and the poetic of land. While philosophers contemplate the intrinsic value of the rest of nature, teachers re-orient learning toward a sustainable future.

Contemporary environmental problems are a seamless mesh of multifaceted issues including: global warming, consumerism, gender and race oppression, resource and biodiversity reduction, globalization, corporatization, drought, knowledge, air and water pollution, poverty, gene patenting, acid rain, urban sprawl, cultural homogenization and ecological imperialism, economic development, land rights, spirituality, and physical, emotional, and psychological illnesses linked to the environment. No one on this planet is immune to these problems. However, it must be understood that the degree to which they are experienced on a daily basis is very much a matter of ethnicity, class, gender, culture, and location.

The contemporary challenge for adult educators worldwide was to recognize and articulate ecologically focussed values, knowledges, issues, and strategies and weave these into the discourse, theory, and practice of adult education. If adult education were to be useful in terms of dealing with interwoven social, cultural, political, economic, and environmental problems, then its theories, principles, and methods would need to be re-conceptualized. It had to become "environmental and just in its essence, spirit and practice" (Benavides 1992, 43). It had to go beyond simply 'greening' adult education activities and programmes, beyond adding environmental issues and stirring. It had to be based on a broader eco-political conception of the world.

Environmental Adult Education

[Environmental adult education] is a process which enables human beings and societies to reach their fullest potential in order that they might live in harmony with themselves and in nature. We also believe that [it] is critical for promoting equitable and environmentally sustainable development and improving the capacity of people, nations and countries to address environment and development issues. (Benavides 1992, 43)

Environmental adult education, or environmental popular education as it is referred to in Latin America and some parts of Asia, is the newest stream of adult education and learning. Although indigenous peoples, women, and others have been practising forms of environmental adult and intergenerational education for centuries, there has been little documentation on their practices or attempts to theorize what this might mean in terms of environmental change.

As the chapters in this book illustrate, environmental adult education is neither a prescribed set of techniques nor a doctrine. It is a collection of principles, frameworks, and critical and creative methods carried out within complex historical experiences and contemporary contexts that are experienced at personal and collective levels.

Contexts of Environmental Adult Education

- who is benefitting from cash crop cultures
- why is there no talk of Shell Oil and Starbucks
- where is the collective accountability of commercial enterprise
- why should globalization mean poor people in Kenya go without rice because of a war in Iraq. (Cole 1998, 103)

Contexts are shaped by cultural and biological diversity, language, place, and human imagination. They are also shaped by capitalism, globalization, corporatization, patriarchy, cultural imperialism, and processes of economic development. By working within these diverse contexts and with multi-layered issues, environmental adult education provides the opportunity to critically examine the root causes of the environmental crisis, calling for analyses that integrate class, gender, ethnicity, and location, the local and the global, the natural and the socio-cultural. Environmental adult educators explore the ways in which these ideologies and processes impact upon human/Earth relationships and variously transform, for good or ill, cultures, ways of living, memory, imagination, values, and creativity.

Many contemporary environmental problems "can be attributed to present development paradigms based in a global economic system" that is neither just nor fair (ASPBAE 1995, 2). These paradigms, built on the ideology of capitalism and globalization, operate at odds with the rhythms of both nature and the majority of the world's people. "The conflict between ecology and economy, indeed between conservation and development, is one resounding reality of our society today. The struggle for development has left a trail of eco-crises" (Usang 1992, 50). Economic development exhausts the supply of natural resources and endangers the survival of indigenous groups and subsistence farmers, most of whom are women.

Greater development through globalization is a concept that has become entrenched in the vocabulary and lives of people worldwide. For some this expansion and 'modernization' has been very fruitful. It has meant an opening up of world markets in terms of a 'freer' form of trade. Freer trade means the ability to export commodities and services as well as ideas, beliefs, and cultures, without restraint. It has made a few people very rich and some better off. But the repercussions for the vast majority, and in particular for women and indigenous peoples, have not been good. The result has been a shirking of responsibility by governments, environmental degradation, greater social unrest and immigration, and growing unemployment, underemployment, and poverty.

Many challenge the belief that there is a positive connection between economic growth and the alleviation of poverty. The African Regional NGOs Think-Tank Consultation Committee (1992, 88) argues that many of the social and environmental ills facing Africa today can be attributed to "the present nature of ownership, control and distribution of African resources." It is "wanton development through savage capitalistic methods of production and commercialisation" that poisons food with "toxic agricultural pesticides" and "contaminates oceans through petroleum exploration" (Viezzer 1992, 4).

Development and the excessive exploration for natural resources have also brought social and economic impoverishment to entire communities. As rural peasants and farmers are moved and/or pushed off fertile lands onto marginal ones, social infrastructures break down (Ibikunle-Johnson and Rugumayo 1987). Thousands of people who made small things in every village in India have been taken over by big companies. "There is less and less creativity. Work becomes mindless. People lose their autonomy. There is increasing alienation, frustration" (Bhasin 1992, 29).

It is important to acknowledge the gendered side to economic development and global restructuring. Decreased expenditures on education, health, and food subsidies mean many increased costs are borne by women. They must work longer hours, take more time to shop for less expensive food, spend more resources and/or time on basic health-care, and face fewer job and educational opportunities. In most communities around the world it is the women who are

> the primary collectors of wood; they draw and carry water for domestic use; they grow and collect most of the food not provided institutionally; they do most of the cooking; they are usually the custodians of traditional knowledge. Therefore, they suffer disproportionately from the effects of environmental degradation and neglect. (LEAP/INFORSE 1997, 5)

In addition, women are still excluded from decision-making levels in the vast majority of government offices, institutions, and organizations.

Development must take place "in terms of 'self' and community change rather than economic growth" (African Regional NGOs Think-Tank Consultation Committee 1992, 88). It must be based on creating internal or local markets which are socially, culturally, and environmentally responsible, and protect them "from external and stronger economies." It must be based on principles of a new partnership with the rest of nature, gender, and racial equity; "creativity, potential and satisfaction" (Bhasin 1992, 32).

James Draper (1976, 69-76) suggests that "the process of social change as a leading edge of development is inevitably linked with the processes of education and learning; education is a vital component in development." For environmental adult education, the broader context of development includes people and ecosystem health, peace, equity, and democracy. For this, we need to push back the boundaries of the purely economic and re-conceptualize development as a life-centred practice of socio-environmental growth.

Ecological Knowledge and the Politics of Education

Peter Sutton (1989, 7) asks, "If we educators wish to teach [people] to be aware of environmental concerns, what precisely can we tell them?"

For environmental adult educators the question is not 'how can we *tell*' but rather, how can we *provide opportunities* for people to draw on their experiences and knowledge, imagine, and work toward a more whole life-centred way of being on this planet.

Although the deficit notion of the 'unaware' public that needs to be 'told' is pervasive in environmental education, environmental adult education begins from a platform of 'ecological' knowledge. As we do not assume that every individual knows everything, nor that, given the complexity and constant flux of the world, there is not always more to be learned. However, we do believe that people have an existing knowledge upon which they can draw. Environmental adult education recognizes that, for the vast majority of people, the concern for the planet remains high. Their knowledge of the gravity, scope, and root causes of environmental problems and their relations to politics and global economics has greatly expanded.[1]

Environmental adult education also moves beyond the 'behaviour' change agenda that has coloured so much thinking in environmental education. We realize that governments who want to shirk responsibility for environmental destruction by supporting multi-national and other polluting industries would prefer that educators work within a framework of 'individual responsibility' and 'behaviour modification' rather than using education to develop a critical consciousness that challenges systemic and structure problems, and re-democratise the political process. Individual behaviour change is insufficient because it can be so easily undermined and the problems are more complex and embedded in destructive ideological frameworks of capitalism and globalization.

Knowing is an ongoing creative and dynamic act which is fundamental to the process of living (Guevara 1995). Environmental adult education also seeks to understand and promote the complexity of ecological knowledge of women and men indigenous peoples, farmers, and fishers. These knowledges or 'ways of knowing' are not esoteric means of comprehending complexity from a different epistemological basis. Ecological knowledge comes from daily lived experience and is based on a comprehensive understanding of a variety of co-varying environmental features and changes over long periods of time. As Sean Kane argues,

> wisdom about nature, that wisdom heard and told in animated pattern, that pattern rendered in such a way as to preserve a place whole and sacred...these are the concepts with which to begin. (1998, 5)

Coupled with more progressive and innovative ideas of contemporary scientists such as Vandana Shiva, these knowledges fill in many of the blanks, helping with what is unknown and perhaps even unknowable. No single group has the monopoly on truth—and indigenous communities, women, farmers, and others who work and live closely with the land have a great deal to offer social development from their time-tested approaches to sustainable resource use, and in creating community. Through ecological paradigms, the recovery, creation or re-creation, and distribution of people's ecological knowledge can become an important contributor to building just and healthy societies (Clover, Follen, and Hall 2000).

Culture and Place

Cultural processes are often intimately linked to the rest of nature. They are much more than mere self-expression or a decorative pastime. They are about identity (Griffiths 1997). As Peter Cole (1998, 104) notes in his poetic article,

> the Pueblo man speaks of the world we bring with us
> he speaks of the relation with one another through ecology
> of having an ecologically specific identity.

For many, the land is culture and culture is embedded in the land, and this relationship is at the root of activities to preserve its value as a source of material existence and spiritual life. In Sri Lanka "a clean and beautiful environment is believed to be both right and a cultural artifact" (LEAP/Ecologic, 1994, 27). In India the forest plays not simply a physical role in sustaining people's lives, but also an emotional and spiritual role, particularly in the lives of women. Women use the forest as the basis of their songs. In Canada, The Group of Seven artists used Canada's landscape as a medium through which to discuss and display an emerging culture. In Newfoundland, music links lives to the sea, while in other places, dance is grounded in the mysteries and ordinariness of the everyday, and the complex symbiotic relationship between humans and place.

Using cultural practices such as mural-making, video production, storytelling, singing, popular theatre, and study tours, environmental adult education challenges dominant forms of education and makes imaginative and aesthetic pathways of learning possible. Cultural forms of representation serve to filter, organize, and convert experience into meaning, transforming the contents of consciousness into a collective form that can be shared (Diamond and Mullen 1999). These can be powerful tools to motivate people to act to save our embattled planet and ourselves.

'Place' is an important element in people's lives. Marla Guppy (1999, 15) argues that place can be "a powerful site in the overwhelming landscapes of globalism." Therefore, learning should not be confined to the indoors or pedagogical institutions but also take place in communities, "in the villages...where people plant, eat, work and celebrate" (Ibikunle-Johnson and Rugumayo 1987, 10). The rest of nature and the community as teachers and sites of learning stimulate ideas, interest, and reflection on the positives of the world, but also the negatives such as poverty, inequality, and racism

(Lengwati 1995). Learning through, with, in, and about the rest of nature can help those who feel alienated to "recover the positive sides and senses of life, and capture its enthusiasm and pleasure...to play, and to entertain" (Vio Grossi 1995, 40). "Learning in and about place" is about the knowledge of place and its network of meanings, of where and who we are: it is about resistance, regeneration, inspiration, beauty, and sensuality.

Activism

As an integration of ideas and practices from adult education and environmental movements discourses and actions, environmental adult education is an activist-based political pedagogy. As such, it goes beyond the notions of individual behavioural change and action. We can practice the 3Rs (recycle, re-use, and reduce), and turn off the tap when we brush our teeth: the reality is that dramatic and fundamental changes are required in oppressive relationships based on gender, ethnicity, sexuality, and class, and in "economic and political structures at national and international levels" (Mikkelson 1992, 72). These changes can be made through the spheres of knowledge, spirit, conscience, emotion, and collective political action.

The problematic of 'action' is that it is either narrowly defined as actually 'manually' doing something, such as tree planting, or it is viewed as something that occurs outside normal daily life. But action or activism are not simply brief, ecstatic outbursts of role reversals. They are personal and collective expressions, formed around contexts, needs, and questions of who we are and where we want to go. Within this framework, educators are able to help people to identify options, making room for the multiplicity of actions that will be uncovered through the educational process. Environmental adult education works to help people begin to include active political expression into their everyday perceptions, work, and lives. This requires an ability by the educator to expand his or her notions of what is often considered 'action' and recognize the daily opportunities presented, through which politics can be expressed and used toward change. It also requires a learning culture of action and mobilization that encourages people to actually see themselves as actors and agents of change.

Creating a "strengthened participatory democracy" Mikkelson 1992) means developing stronger cross-cultural and intersectoral networks; partnerships amongst educators, environmental, and conservation groups, citizen organizations, academics, and members of other social movements. It also provides a conceptual framework within which to render visible and explore the subtle (and often not so subtle) race, gender, and class biases found in environmental activism and advocacy.

Our Stories

This book contains twelve chapters from Australia, Canada, Fiji, India, Kenya, Mexico, Philippines, Sudan, and the United States. The stories are divided into three main sections. Although there is much inter-weaving of ideas throughout the texts, the sections focus us on traditional knowledge, women's learning and activism, and practice and research. The authors do not attempt to offer a comprehensive overview of environmental adult education but, rather, they have the modest ambition of illustrating some of the contexts, issues, practices, and principles from diverse parts of the world.

Although the contributors come from a variety of locations, both inside and outside the academy, they are primarily practitioners who, by working on the ground, have something significant to contribute to the enrichment of the epistemology and pedagogy of human-Earth relations. Their collective work illustrates how the development of educational processes is a social process that emerges from the interaction of action and thought in specific settings and contexts.

In chapter one, Sandra Tan analyzes the racialization of the northern environmental movement and how immigrant communities from the south participate and negotiate their place within a white-dominated movement. She traces the entrenchment of white subjects in the movement to the historical processes of European expansion and imperial relations that created formalized systems of knowledge about the natural world, and bestowed authority upon these subjects in discoursing the environment. Colonial relations also produce racist ideologies and unequal relations which are re-articulated in environmental discourses. Through a study of the work of two immigrant environmental organizations in Toronto, she challenges the 'popular' perception of the 'ignorant' immigrant who is unaware and unconcerned about environmental problems.

Rosa Muraguri-Mwololo in chapter two illustrates how past colonialism and contemporary development initiatives have managed to undermine the traditional ecological knowledge of pastoralists who roamed the desert of northern Kenya, thriving within its harsh environment for centuries. Through an examination of a participatory educational initiative by a community-based organization to revitalize traditional knowledge of pastoralist management systems, Muraguri-Mwololo illustrates how critical environmental adult education can weave the old and the new, awakening dormant traditional ecological knowledge to cope with contemporary problems.

Dip Kapoor examines environmental popular education activities with the Kondh *adivasis* of Orissa State, India, in chapter three. He traces the attempt to democratize entrenched structures and practices of power which marginalize hundreds of people and the social theories that underpin these endeavours. Framed within the discourse of new social movements and new theoretical assumptions about indigenous knowledge and rights, Kapoor examines why environmental popular education, with its synergy between environmental concerns and its focus on the question of disempowerment, silence, and cultural identity, has been able to find a place within the *adivasis*.

In chapter four, Darlene E. Clover examines the coming together of ecology, feminist leaning, and education and aesthetics. Through a feminist environmental adult education lens, she explores the ways in which women artist-educators in two diverse locations in Canada—the urban and the rural—use crafts or visual arts to mobilize, revitalize, challenge, and transform notions of 'place.' The use of aesthetic activism and learning provides new paradigms to comprehend and value art, and to use it to promote socio-ecological consciousness, and to challenge the status quo.

Salwa B. Tabiedi explores some of the basic elements of environmental adult education with, for, and about women in Sudan in chapter five. Through an examination of the work of the Science Sector of the Sudanese National Commission for UNESCO, Tabiedi highlights a variety of activities such as seminars on environmental health, tree planting, workshops with urban slum dwellers, media campaigns, and the interdependence between literacy and power. She concludes with a number of the challenges environmental adult educators in Sudan face and how these can be overcome through respect for women's knowledge, and their active participation in learning.

Kerrie Strathy and Kesaia Tabunakawai explore an environmental adult education project for women in Fiji, in chapter six. Their case study is an innovative environmental adult education workshop entitled "Women and Forests." This workshop took place in the forest where women could better learn with and through the rest of nature about everything from the location and use of traditional medicinal plants to the devastation of logging. The work generated a video, new work on the issue of intellectual property rights and indigenous knowledge, recognition of the importance of traditional medicine, and greater respect for the knowledge of women healers.

Joaquín Esteva and Javier Reyes examine the multi-faceted environmental popular education work of the Centre for Social and Ecological Studies in Mexico, in chapter seven. The programmes include participatory strategies of research, citizenship building, community health, and organizational development for women. The authors argue that environmental popular education cannot restrict itself to merely passing along ecological information, but rather must draw out people's embedded environmental consciousness, and develop collective ethical ways of thinking that enrich people's lives. A major lesson is that, without the full participation of women in learning and social change activities, the programmes are doomed to fail.

In chapter eight, José Roberto Guevara of the Philippines examines the transformative nature of environmental adult education from four different perspectives. The first is the Philippine environmental movement, which he describes as a story waiting to be transformed, and the importance of making connections between education and activism. The second is a series of workshops and their contribution to shaping the theory and practice of environmental adult education. The third and fourth respectively are Guevara's own story of transformation as an educator and learner, and the field of environmental adult education as a whole, in terms of its role in socio-environmental transformation.

In chapter nine, Karen Malone examines fundamental notions of "who has knowledge" and "who the educators are." Through two case studies, she examines two diverse views or approaches to environmental learning: the information sharing/behavioural change paradigm, and the social transformation process of environmental adult education, relating the latter to the oppositional ideology of the Australian environmental movement. Malone argues that environmental activism is an important form of environmental adult education, and the activist, an important educator.

In chapter ten, Jan Woodhouse analyzes three participatory action research (PAR) projects in various parts of the world in order to highlight the characteristics of environmental adult education. She examines the theoretical foundations of knowledge formation, the processes in which the adult educators engage, and the various principles of PAR that apply to learning and citizen action. This illumination highlights environmental adult education as a tool for change, moving it from the margins of adult education to the centre of contemporary discourse.

Budd L. Hall shares the findings of an international research study titled "Transformative Learning Through Environmental Action" in chapter eleven. The study examined the ways in which learning emerged from seven environmental campaigns from Brazil, Canada (Six Nations Aboriginal Community), El Salvador, Germany, India, and Venezuela. The first of its kind at the international level, the study highlights key principles of a transformative environmental adult education such as the recovery of a sense of place, acting and resisting and deconstructing power as it emerged through case studies, the research process, and a workshop.

In the final chapter, Darlene E. Clover and Shirley Follen reflect on a community-based process to develop the first practice-based environmental adult education resource. Given that the vast majority of environmental education resources have been designed for children and teachers in schools, Clover and Follen worked together with community groups and adult educators across Canada to develop critical and creative practices for working with adults in an ecological context. Their candid examination of the successes and failures of the work reveals the slow, painful, yet fulfilling nature of critical environmental adult education and community change.

Note

1   From The Globe and Mail newspaper (Toronto) during the June 1997 'Mini-Earth Summit' at the United Nations headquarters, page A2.

References

African Regional NGOs Think-Tank Consultation Committee. 1992. "A Common Position: UNCED and Beyond." *Convergence XXV*, 2, 88–92.

ASPBAE. 1995. "Evolving an Asian-South Pacific Framework for Adult and Community Environmental Education." Environmental Education Programme. Darwin: ASPBAE.

Bhasin, Kamla. 1992. "Alternative and Sustainable Development." *Convergence XXV*, 2, 26–36.

Benavides, Marta. 1992. "Lessons from 500 Years of a "New World Order"—Towards the 21st Century: Education for Quality of Life." *Convergence XXV*, 2, 37–45.

Clover, Darlene, Follen, Shirley, and Hall, Budd L. 2000. *The Nature of Transformation: Environmental Adult Education*. Toronto: OISE, University of Toronto.

Cole, Peter. 1998. "An Academic Take on Indigenous Traditions and Ecology." *Canadian Journal of Environmental Education 3*, 100–115.

Draper, James. 1976. "The Significance of Habitat 1976 to Adult Education." *Convergence IX,* 4, 69–76.

Diamond, C.T. Patrick and Mullen, Carol A. 1999. The Postmodern Educator: Arts-Based Inquires and Teacher Development. Peter Lang, New York.

Elsey, Barry. 1986. *Social Theory Perspectives on Adult Education.* Nottingham: Department of Adult Education, University of Nottingham.

Grace, Andre. 1998. "Parameters, Pedagogy and Possibilities in Changing Times." In Sue Scott, Bruce Spenser, and Alan Thomas (Eds.), *Learning for Life: Canadian Readings in Adult Education.* Toronto: Thompson Educational Publishing Inc.

Griffiths, Jay. 1997. "Art as Weapon of Protest." *Resurgence 180, January/February*, 35–37.

Guevara, José Roberto. 1995. *Renewing Renew: A Restoration Ecology Workshop Manual.* Manila: Centre for Environmental Concerns.

Guppy, Marla. 1999. The Edge of Centre: Local Identity and the New Neighbourhood. *Network News, Edition 3*, 9–15.

Ibikunle-Johnson, Victor and Rugumayo, Edward. 1987. *Environmental Education Through Adult Education.* Nairobi: African Association for Literacy and Adult Education.

Kane, Sean. 1998. *Wisdom of the Mythtellers.* Peterborough, Ontario: Broadview Press.

LEAP/Ecologic. 1994. Adult Environmental Education: A Workbook to Move from Words to Action. Toronto: ICAE.

LEAP/INFORSE. 1997. Working Conceptual Paper on Environmental Adult and Popular Education. Toronto: OISE, University of Toronto.

Lengwati, Makkies David. 1995. "The Politics of Environmental Destruction and the Use of Nature as Teacher and Site of Learning." *Convergence XXVIII,* 4, 99–105.

Luke, Carmen and Gore, Jennifer. (Eds.). 1992. *Feminisms and Critical Pedagogy.* London: Routledge.

Mané, Albert Martin. 1993. "It is Time to Understand that the World Belongs to All of Us: Problems of the European-African Dialogue." *Adult Education and Development, 40*, 7–12.

Mikkelson, Kent. 1992. "Environmental and Adult Education—Towards a Danish Dimension." *Convergence XXV,* 2, 71–74.

NGO FORUM/LEAP. 1992. "Treaty on Environmental Education for Sustainable Societies and Global Responsibility." Rio de Janeiro, NGO Forum.

Reason, Peter. 1998. "A Participatory World." *Resurgence 186, January/February,* 42–44.

Rosero, Rocio. 1993. "The Challenges of the Democratisation of Popular Education from the Perspective of Gender." *Convergence 26,* 1, 80.

Selman, Gordon and Dampier, Paul. 1991 and 1998. *The Foundations of Adult Education in Canada.* Toronto: Thompson Educational Publishing Ltd.

Sutton, Peter. 1989. "Environmental Education: What Can We Teach?" *Convergence XXII,* 4, 5–12.

Thomas, Alan. 1991. "Relationships and Political Science." *Adult Education, Evolution and Achievements in a Developing Field of Study.* Oxford: Peters, Jarvis and Associates, Jossey-Bass Publishers.

UNESCO. 1976. Recommendation on the Development of Adult Education. Paris: UNESCO.

Usang, Ewa A. 1992. "Strategies for Green Literacy." *Convergence XXV,* 2, 46–54.

Viezzer, Moema. 1992. "A Feminist Approach to Environmental Education." Paper presented
    at World Congress for Education and Communication on Environment and Development
    (ECO-ED), Toronto, October.
Vio Grossi, Fransisco. 1995. "Cambios Paradigmáticos Sociedad Ecologica y Educación."
    Convergence, XXVIII 4, 31–43.
Welton, Michael. (Ed.). 1995. *In Defense of the Lifeworld: Critical Perspectives on Adult
    Learning*. New York: State University of New York (Albany).

Part I

Traditional Ecological Knowledge and the Issue of Race

Chapter 1

Anti-Racist Environmental Adult Education in a Trans-Global Community: Case Studies from Toronto

Sandra Tan
Toronto, Ontario

Revisiting the Empire

Discourses about 'nature' have long been dominated by Northern scholars and academics. The origin of this monopoly can be traced back to Europe where the history of modern ecology and environmental science was given existence, in the name of the empire. Mary Louise Pratt (1992) has traced the emergence of a *planetary consciousness* with the rise of environmental science in the era of colonialism. This alignment helped forge the European consciousness of their existence in relation to the rest of the world, a consciousness characterized by a desire to impose a planetary ordering on people and the universe, thus rendering it a "basic element constructing modern Eurocentrism" (Pratt 1992, 15).

Mercantilist expansion and territorial rivalry among European powers in the seventeenth century saw the deployment of members of their male bourgeoisie to the interior territories of places beyond the European world. The ahistorical, asocial, objective, and scientific 'knowledge' gleaned from the imperious gaze of these 'gentlemen travelers' became the bedrock of dominant discourses of nature, ecology, economy, anthropology, biology, history, etc. (see Pratt 1992; Goldberg 1993).

In recent years, globalization has caused massive changes in social, political, economic, and ecological conditions that, in turn, have been catalysts for new patterns of migration, often on such a scale as to re-configure national politics and cultures. For example, the flow of new immigrants into Canada each year has reshaped what was hitherto understood as the essential 'Canadian' identity. Coupled with growing problems such as unemployment— primarily due to globalization— this has fueled new manifestations of racism in Canada. This chapter examines one of these manifestations—the ways in which traditional nature-based environmental education for adults in Canada tended to ignore the issues of race. Through an examination of two case studies from Toronto, I analyze the ways in which the contemporary theory and practice of environmental adult education problematizes race and works to incorporate the ecological perspectives, interests, and knowledge of a diverse citizenry.

Mapping the Web of Environmental 'Problems'

The historical entrenchment of white subjects as the discursive authority in the study of the rest of nature is reflected in the polarized North-South debates on the global environment today. Many Northern environmentalists tend to view the roots of ecological problems and the onus for change as lying outside of themselves and their context. The South, they argue, is the source of ecological destruction through an endless list of calamities including overpopulation, deforestation, desertification, disappearance of wildlife, pollution, hunger, and disease (Shiva 1994). Such views gain validation when they are reiterated by institutions of global authority such as the World Bank. Even the United Nations Conference on Environment and Development, Vandana Shiva observes, saw the South "as the source of all environmental problems and the North, with its technology and capital, as the source of all environmental solutions" (2).

These views have had a strong influence on dominant Northern interpretations of diasporic responses to the environment in the South. As with many environmental issues, population discourses are often politically motivated (Poore 1993). They are also controversial as they impinge upon national, racial, ethnic, class, and gender ideologies. As a result, they are prone to subversion and may be invoked in the service of fascism and other agendas of domination. A particularly disquieting example of this was the attack on Latin American immigration to the United States of America by Dave Foreman of Earth First! Foreman argued that "it's just putting more pressure on the resources in the USA" (in Merchant 1992, 175). Comments such as these reflect the way environmental issues are so often "presented from a single point of view" and with "little historical perspective" (Poore 1993, 29).

Connecting Race and Traditional Approaches
to Environmental Education for Adults

> It has been my experience that most of you aren't going to deal with the problem at
> the level that it is going to help the welfare recipient, the poor person in the ghettos
> and the barrios. Most of you are not even going to listen to the voices coming from
> those communities. You won't ask what they want, and how they want to deal with
> the problems of their environment, or indeed whether they want to deal with the
> problems of environment at all because they feel there are other more pressing
> priorities in their lives. (Wiley 1994)

Several studies have noted the low representation of people of colour in the
Northern environmental movement, being dominated as it by those who are
educated, white, and middle-class (Shabecoff 1990; Lewis and James 1995;
Lichterman 1995; Newman 1994; Pickering 1995). For example, Lichterman
(1995, 513) argues that environmentalists in the United States have
"accomplished relatively little in their multicultural alliance-building quest,
and appeared fated to remain a largely white, [and] highly educated middle-
class group." Pickering specifically pointed to the failure of environmental
groups in Canada to establish alliances with immigrant communities. Within
this framework, environmental issues are often projected as universal
concerns demanding universal attention. Such a narrow, bourgeois view
precludes the voices and environmental aspirations of those who struggle
against unequal relations and systems of oppression. They have also led to
the simplistic conclusion that people of colour are less concerned about the
environment than white people (Lewis and James 1995).

Sly (1991, 2) noted the educational consequences of this when he
contrasted the way the environmental education practiced by many in the
environmental movement tends to emphasize "the values and lifestyles of
white, middle-class students" while "environmental issues impact all races
and all socio-economic groups." As a result, the only forms of
environmental education for adults—better referred to as outdoor
education—in the past in Canada tended to involve wilderness exploration
and other such activities. And these tended to be the pursuits of those who
are mostly male (Clover 1998; Watson et al. 1992), white, middle-class, and
well educated (Bultena and Field 1978; Wilson 1991; Sly 1991). This
reflects the fact that environmental programme developers also often hail
from this particular social group (Gibson and Moriah 1980). Even where
there have been attempts to examine human impacts on the natural
environment, these have tended to focus on the preservation of wildlife and

ecosystems. Rare are the cases where they have engaged with the socio-economic problems that are manifested in the deterioration of living and working conditions, social disenfranchisement, and urban atrophy (Gigliotti 1990). Baker (1991, 5) noted the inherent gaps in these programmes when he lamented that he "could find few environmental education materials appropriate to a multi-cultural, urban setting."

A significant problem that results from this may be seen in the way in which dominant discourses on environmental thought treat issues of power as dichotomous categories between 'society' and 'nature.' This view sees society as universal, homogenous, and monolithic, and devoid of multiple and competing claims to histories and social-material bases. Nature's oppression is viewed as the result of its enslavement by the human species. Such reductionist theorizing becomes dangerous when its ambiguities are left unquestioned, and especially when it neglects the transnational dynamics of race, class, and gender in explanations of global environmental degradation (Livingston 1994; Potts 1996). While environmental degradation occurs across the globe in different geographical spaces and degrees, the lack of critical and contextualized analysis distorts interpretations of environmental problems, and subverts efforts to situate the real causes of the environmental crisis (Bookchin 1995; Seager 1993; Shiva, 1994). This is similar to the way that the generic categorization of the 'human species' subsumes the centrality of power deriving from unequal distribution of social-economic and political entitlement among dominant and subordinated groups. This has the effect of not only submerging the environmental discourses of marginalized groups, but also the dissimilarities among them, thus rendering the perception that people of colour are invisible in the environmental landscape (Lynch 1995). Since communities and society are affected by the complex interconnections of geographical, cultural, social, material, and political processes, they are experiencing environmental problems and reforms differently and respond to them in nuanced ways. Luanne Armstrong (1993, 6) notes the interconnectedness of social, political, and material relations when she writes that:

> exploitation, abuse, and degradation of the environment, the exploitation of cheap labour through colonization, slavery, racism, poverty and systematized oppression, and the gender oppression inherent in the social, economic, and political structures of women's lives, are crucially interrelated.

Hence, the nature-culture dualism cannot be reconciled without attention to the patterns of social domination that often go hand in hand with the exploitation of nature (Sen 1994). This means that concern for human well-being should not necessarily be taken as an anthropocentric, rather than eco-centric, ethic. Instead, it gestures to an affirmation of the inseparable boundaries between humanity and the natural world that sustains us (Lewis and James 1995; Shiva 1994).

Locating 'Immigrants' in the Discourses of Environmentalism

To appreciate the absence of immigrants from the discursive landscape of environmentalism, it is necessary to examine some of the historical and contemporary issues that link discourses of the environment with those of immigration and nationalism. Canada is one of the few countries that continues to accept refugees and immigrants, making it one of the most culturally diverse countries in the world. Toronto is considered to be the most multicultural city in the world. However, this contemporary facade of benevolence, tolerance, and liberalism masks a historical chronicle of repression, discrimination, harassment, coercion, and exclusion. Many forms of these practices were inflicted on various groups of people along racial lines at numerous historical junctures. For example, the period of colonial expansion saw the exploitative and oppressive treatment of non-white immigrants as a source of cheap labour necessary for the development of the western wilderness. White European immigrants, on the other hand, were sought as farmers to "break, till and sow the soil" (Elliot and Fleras 1990, 54). Inevitably, these circumstances defined the relationships with the land of white and non-white immigrants. The processes of conquest and settlement fostered a renewed sense of home and belonging to the new land in the former, while the latter experienced much of the New World as transient migrants whose presence was viewed with suspicion and aversion.

Despite the significant contributions of non-European immigrants to Canada, the majority of Canadians harbour deep-seated resentment toward immigrants (Elliot and Fleras 1990; Miller 1997). Popular ideologies of racism and xenophobia invoke convoluted myths of immigrants 'flocking into the country' either to 'make money' or to escape the brutality and barbarism of their war-torn, underdeveloped, and culturally oppressive homelands. At the same time, those who braved the perils of flight in search of a Canadian refuge from violence and oppression are labeled as 'frauds,' deemed 'unfit,' and refused a ready welcome to Canadian society. The knowledge systems and experiences of new settlers, including their ecological praxis, find little legitimacy or are dislodged altogether amidst such racist refigurations. Crude ideas of immigrant ignorance gained normalcy from colonial representations of the Other and within the discursive formation that comprised the Northern consciousness of ecology. The recasting of peripheral peoples as lacking culture and consciousness is evident in the way they are projected in popular environmental discourses. John Livingston's *Fallacy of Wildlife Conservation* (1981, 20) illustrates this point:

> In underprivileged countries, I see no glimmer of hope for wildlife. It is simply too much to expect, for an entire catalogue of reasons, that the care and maintenance of wildlife could possibly rise on the list of priorities (including accelerated industrial expansion) that exists in the tropics today. We should remember that there is *little or no preservation tradition in most such places*, in any event, and to think that such a tradition could spring forth fully developed in the face of current events would be to abdicate common sense altogether. (my emphasis)

Here, the colonial discourse of a forlorn Third World converges with the discourse of Northern enlightenment to construct an essentialized binary, positioning Northern environmentalists as enlightened saviours against the ecological villains of the South. The authority of this discourse is made lucid by the absence of any examples from the 'catalogue of reasons' on which the fundamental premise of the text is founded. The broad assumption is that people in the tropics (who ironically have experienced most of the violence connected to imperialism) are too conceptually ignorant and culturally limited to embrace an ecological tradition like those from 'civilized' parts of the world. Tacitly contained in this omission is the insinuation that the 'catalogue of reasons' is simply not worthy of elaboration since it is assumed that the context would be spontaneously apparent to a rational and enlightened First World audience, thus reaffirming white hegemony and authority in knowledge production. What should be noted is that the people in 'underprivileged countries,' plagued as they are by a profusion of crises, are simply incapable of making wilderness preservation a priority.

Even though the *Fallacy* was published almost two decades ago, the assumptions contained within have persisted. For example, in 1995 the director of the Vancouver Aquarium reiterated Livingston's assertions by singling out 'immigrants' as the perpetrators of illegal fishing in Stanley Park. Under projected slide shots of an Asian family harvesting clams at Whytecliffe Park in Vancouver, allegedly ignoring a prohibition sign, he remarked, "Cultures unfamiliar with the notion of conservation are devastating Stanley Park's marine life.... [I]mmigrants from different cultures have the attitude: 'If it's alive, eat it.' These people come to Canada and they need better education" (Pynn 1995, A1). The argument that immigrants require 'better education' reinscribes the primacy of Northern enlightenment and the trope of immigrant ignorance. Despite evidence to the contrary, the speaker who presented the slides insists that the family in question "knew" their activities were unlawful, but simply "didn't care" (Pynn 1995, A1). Given this supposition, interpretations of resistance are denied possibility, if not silenced, by this juxtaposition of consciousness with indifference. A second example suggests an alternative reading. Research with working-class Puerto Rican immigrants in the United States shows that they do not resist restrictions on recreational fishing in public waters out of ignorance or villainy. Rather, recreational fishing embodies a cultural significance built into family life, while sharing marine harvests with friends and neighbours affirms bonds of community. Thus, they see restrictions on fishing as infringing upon their freedom to interact with nature and to recuperate nostalgic expressions of sharing and community in the diaspora (Griffith et al. 1992).

The salience of race and ethnicity in constructing responses to divergent environmental perspectives also surfaced in other studies. For example, Cunningham (1998, 25) reports a study by three academics that "godless Caucasian females are the strongest environmentalists" in British Columbia in contrast to "men or Asians." And to be sure that the term 'Caucasian' forestalled any misinterpretation, the report further delineated the specificity of this whiteness by locating its nexus in "white women" of "European origin" (25). A second example is found in a study of the environmental attitudes of environmental club members in two public schools in Toronto (Lousley 1998). While this research focussed on the way students express their environmentalism, conversations with teachers revealed a racialized perception of student environmentalism. When asked why black students of urban working-class roots were less enthusiastic about a wilderness experience than white middle-class students, the white educator "resorted to a 'savages in the urban wilderness' explanation for their inability to, in his words, 'connect with nature'" (Lousley 1998, 179). On the contrary, the idea that black youths are equally capable of an appreciation of nature and its preservation rattles popular conceptions of who is associated with environmental activism. In this example, a team of African-Canadian youths engaging in a tree-planting project in the Don River Valley, on a hot summer day in Toronto, was mistaken for labouring convicts by the largely white recreational visitors to the park.

Debunking the Myths of the Ignorant Immigrant

Both Northern communities of colour and Third World peoples have been actively defining their own ecological struggles despite assertions that immigrants are bereft of ecological traditions. Often this has involved engaging in militant resistance to the imposition of ecologically unsustainable development (see Day and Knight 1991; Newman 1994; Shiva 1994; Banuri and Marglin 1993; Omvedt 1994; Agarwal 1992).

For many women from the Third World, ecological praxis comes primarily from their productive labour in managing, harvesting, collecting, and processing food, water, and fuel. My own research has been primarily with immigrant women in Canada involved in environmental adult education (Tan 1999). Interviews with these women show that they practice ecological traditions from their native homelands. These practices arise as responses to dealing with the daily struggle with health, basic needs, and survival issues, elements that differentiate Third World ecology movements from those in the North (Chambers 1987; Shiva 1994). In recounting her upbringing in a Kenyan village, one of the women I spoke with described the ecological habits of her community: "[W]e don't throw away anything. We fix up everything and we mend, and we re-use as much as possible.... [W]e were very, very much trained to conserve and to save." Food scraps and leftovers are never wasted, but fed to household animals, or composted. Their ecological consciousness is not isolated to individual practices, but extends to social relations that underscore the core values of sharing and reciprocity in regulating community life.

Other immigrant women have spoken of using a handkerchief instead of disposable tissues and adopting water conservation strategies they cultivated at home (Gokhale, in Pickering 1995, 14). Asian women environmental educators I interviewed describe how Chinese residents in Toronto are already engaging in environmentally responsible habits in their daily lives, all the while oblivious to the ecological value of their practices. For societies where ecological sustainability traditionally mediates daily life-sustaining activities, the sundry performances of environmentalism by members of such societies are undertaken unconsciously and removed from the formal conceptualization of 'environmentalism.' In this context, the deep sense of frugality demonstrated by the residents and their predisposition to conserving and reusing household resources issue from a historical memory of resource scarcity in the Chinese homeland. There, long-term population pressures and chronic shortages of food and other essentials have compelled the people to devise coping mechanisms in the rituals of everyday life. This explains why sharing is a fundamental aspect of Chinese custom commonly observed in the congeniality of communal meals. It also explains why many Chinese are apprehensive about wasting food. Hence, an animal sacrificed at the altar of gastronomy is always ensured a complete purveyance by the Chinese cook. Every part of its offal is generally put to some consumable use (much to the abhorrence of the western palate!).

In the process of migration, the symbolic origins of these common sense practices are transported and given new expression in the diaspora. Some immigrant women have used their culture-specific ecological knowledge to help other immigrant women maintain environmentally friendly practices while nurturing new ones. For example, immigrant women have run transportation workshops to raise awareness about cars and pollution, and created bicycle repair workshops to promote women's independence and alternative transportation (see Pickering 1995). For the *migr*, the process of rebuilding home and community in a distant land entails reconciling the contradictions of the host country with the indigenous practices of the native homeland. But this shift can be traumatic and overwhelming, leading many to abandon ecologically sensible practices and belief systems cultivated from home. The culture of consumerism underlying Northern economic relations and ideologies of progress are often pitted against communal-oriented and subsistence modes of production of parochial societies, rendering them as incongruent and antithetical to the values of modernity (Pickering 1995; Ralston 1991). In this context, reusing and recycling thus become signifiers of destitution and inferiority, while extravagance and change are viewed as markers of privilege and modernity.

Environmental Adult Education as Resistance

Immigrant communities have responded to their alienation by Northern environmental educators by defining their own approaches to nurturing ecological consciousness. They take into account the challenges confronting immigrants in their daily lives, such as racism, linguistic barriers, unemployment, and poverty. The following case studies will show how two immigrant groups in Toronto have drawn on the socio-cultural particularities within their communities to develop creative innovations and particularistic approaches to traditional environmental education for adults. Since they ascribe to the philosophy that ecological healing comes only through the revitalization of the vision, strength, and moral values of a community, their initiatives embody a narrative of resistance and renewal.

The Toronto Chinese Health Education Committee (TCHEC)

The Toronto Chinese Health Education Committee is a coalition of community and municipal health agencies established in 1985, with the goal of providing health education to the Chinese-speaking community in Toronto. Over time, a growing concern for the environment within the community led to the formation of an environmental sub-committee in 1990—with a mandate to provide environmental adult education in the Chinese vernacular. As an auxiliary arm of TCHEC, the sub-committee is by no means a fully functional environmental group, and is limited by technical and financial difficulties that undermine the extent to which it can effectively carry out its programming.

TCHEC members are primarily community health workers with little or no background in traditional forms of environmental education for adults. Its members also confront a major problem in providing environmental information in languages other than English and French. The ability of the sub-committee to overcome this obstacle is a reflection of its political will and creativity. Margaret Chiu, the resource person of the sub-committee, describes how she and her colleagues extricated mounds of environmental information and materials in English before selecting and painstakingly translating materials of particular concern to the community into Chinese. It was through these uncanny channels, she reminisced, that she and members of the sub-committee have educated themselves and their community members about the environment.

The sub-committee has expanded its work from these humble beginnings, and produced pertinent and much-needed resources in Chinese, including a range of multi-media educational tools that address concerns such as indoor pollution, health hazards, water and energy conservation, composting, and ecosystem issues. In addition, it has also launched environmental projects that are contextually relevant to the community. For example, to bridge community concerns with those of the commercial sectors, the sub-committee initiated an innovative joint project with the Ontario Chinese Restaurant and Food Services Association, to help restaurant owners improve their waste management and energy consumption.

The linguistic and social barriers that keep immigrant communities marginalized and alienated from sources of information and public services led the sub-committee to diversify its modes of community outreach. This has included a 30-second television commercial in Chinese promoting an environmentally responsible lifestyle; community workshops, seminars, and hands-on demonstrations on the use of alternative home products; and a weekly column on environmental adult education in a Chinese daily. The sub-committee also undertakes joint projects with the local community and municipal government to restore and protect green spaces. TCHEC's information kit notes that these initiatives have generated increased awareness of environmental issues. The sustained enthusiasm of the sub-committee led to the establishment of an environmental ambassador programme, in which members volunteer as 'environmental ambassadors' to carry out ongoing surveys of local residents, to identify their environmental concerns and the educational approaches most suited to the community. The findings of the ambassadors are reported to the sub-committee, which then uses the information to plan future projects and education tools.

The modest initiatives of TCHEC's sub-committee represent the kind of grassroots interaction and consultation emphasized by many participatory researchers in which the community identifies its own needs and determines its own means of addressing the problems (Kirby and McKenna 1989; Maguire 1987; Nelson and Wright 1995). Lewis and James (1995) affirm that, where environmental adult education is not readily available and accessible, community outreach and recruitment are important channels of communication.

In the context of social justice, the Sub-Committee has made new theories and practices of environmental adult education available to a segment of the population which would otherwise be excluded by linguistic and other socio-economic barriers. It has enabled the Chinese-speaking community to articulate their concerns and create a space in which they can shape their own learning programmes that speak to their linguistic and cultural particularities. These experiences have enabled one of the participants to challenge the misconception that the Chinese community is not actively engaged in environmental praxis. Instead, she argues that traditional environmental education for adults was often discursive or metaphysical and removed from the grasp of ordinary people. To the contrary, she adds, "we found out that when we talk to people, they are more aware of the environmental issues and the role that they can play." As a result, she argues that western notions of public action should be questioned to reveal their limitations in the face of these diverse expressions of environmentalism. She asks, "If they only work in their house, do you consider that participation? Or would you consider them [the Chinese community] participating only if they come out to join discussion, or voice their opinion, or interest, or question?" Yet, people who have historically been intimidated to express their opinions or dissent, and those who are socially and economically oppressed, may not be comfortable at 'public' sites of participation such as protests, marches, and rallies (Tan 1999).

By incorporating new ideas of environmental adult education into health education, the Sub-Committee is reaching a wider audience and addressing adults from all walks of life, including those who do not speak the dominant languages. The work of the Sub-Committee has been central to the development of ecological literacy within the Chinese-speaking communities in Toronto. It has empowered people to contribute their knowledge and to play a role in enhancing the environmental sustainability of their neighbourhoods. In so doing, the locus of environmental knowledge and decision making is redistributed from the hands of ecological 'experts' into those of ordinary people.

Environmental Centre for New Canadians (ECENECA)

ECENECA, the Environmental Centre for New Canadians was formed in response to the exclusionary practices of established major environmental groups in Toronto. It began with the vision of a Ugandan immigrant, Yuga Onziga, of an environmental foundation that addressed the ecological concerns and social issues confronting immigrants and the underprivileged. As Onziga puts it:

> People who are struggling to get social recognition, social consciousness are left out of the environmental movement and, because of the narrow definition of the environment, the middle-class left out those people who are socially poor and they happen to be people of colour or visible minorities in the case of Canada.

Onziga's own critical consciousness and political awareness spring from his experiences and struggles as an immigrant of colour in Canada. Once an active volunteer in mainstream environmental groups, Onziga encountered opposition to his calls for greater recruitment of immigrants to the field. He says that remarks such as 'we don't have funds [to work with immigrant groups]' or 'these visible minorities are not interested in it [the environment]' were common assertions to justify the invisibility of visible minorities from mainstream environmental organizing.

ECENECA views socio-economic problems as inherently linked with environmental problems, and strives to address the two as interlocking issues. It argues that immigrants are knowledgeable and have much to contribute to Canadian society, but are prevented from doing so by a racist culture bent on portraying immigrants as inferior and as burdens to the state.

In line with this critical awareness, ECENECA works to bring new practices in environmental adult education to marginalized communities across Toronto, with specific emphasis on the African-Canadian and low-income communities. It carries out environmental adult education to a wide spectrum; community development work is also integrated. As an environmental organization, ECENECA is unique; it is not only a storehouse of information on issues locally and globally, it is also a safe and welcoming space for new Canadians to mingle, gain new skills, and share ideas and resources.

The incorporation of a social imperative into the environmental framework expresses ECENECA's validation of the social and economic hardships that new immigrants encounter in securing employment, learning a new language, adjusting to a foreign culture, and confronting everyday racism (see Essed 1991; Miller 1997). It attempts to address social and structural barriers encountered by immigrants by taking a multi-disciplinary approach to developing an ecologically responsible citizenship. Hence, it offers community networking, employment skills seminars, computer literacy and writing workshops alongside tree-planting projects, ecosystem restoration, nature exploration, and seminars on transportation, pollution, energy, and other issues. Socially, it has created environmental literature which is representative of the cultural-linguistic milieu of immigrant communities. Politically, it has challenged the media to produce documentaries on the accomplishments and contribution of black people in the fields of science, technology, and the environment. Onziga relayed to me that the television programme, *Discovery*, responded to this challenge by broadcasting a documentary on black scientists in conjunction with Black History Month. ECENECA also attempts to recognize Canada's multi-cultural mosaic, and strives to make environmental adult education accessible to those who are not proficient in the official languages. The strategies that ECENECA employ reflect the importation of indigenous African and folk values of communal responsibility, co-operation, pragmatism, innovation, and reciprocity into a multi-racial setting of an industrialized metropolitan. In developing its vision toward the development of a progressive philosophy for social change, the group draws on the principle of what Dei terms as 'collective responsibility' (Dei et al. 2000). This principle stresses that one's entitlement and membership within the group must be earned through one's contribution to society.

Reflections

The emergence of minority groups such as TCHEC and ECENECA represents a small but significant step toward the inclusion of immigrant communities in environmental adult education. More significantly, their work helps to shatter the myths that immigrants are ecologically ignorant and lacking agency. It also illuminates the cultural barriers that dominant or traditional models of environmental education for adults, those still most often used by the environmental movement, have yet to transcend. Lichterman (1995) highlights the salience of cultural differences as an impediment to building multicultural alliances among activist groups. In this context, the individualized approach to community building that is practiced by many green activist groups "unwittingly perpetuates class or race divisions" (528). At present, TCHEC and ECENECA and similar groups are ghettoized from the dominant environmental community. Their effectiveness and sustainability are also threatened by chronic shortages of funds and other resources.

Pedagogically, the two case studies cited exemplify what many authors have already argued—that an effective programme in environmental adult education challenges traditional forms of environmental education and is more relevant to the issues affecting the community (Gigliotti 1990; Ham et al. 1985). As Freire (1970, 76) emphasizes, "the starting point for organizing the programme content of education or political action must be the present, existential, concrete situation, reflecting the aspirations of the people." Educational programmes framed by this understanding recognize that the lives of adults and the daily actions they take are affected by interlocking social, material and political relations. Indeed, ecological sustainability must not be viewed as an isolated phenomena and external to these relations, but as inherently linked to issues of survival, health, and social and political justice (Sen 1994; Shiva 1994; Bodley 1990).

Considerations for Change

Critical anti-racist and multicultural educators have provided cogent insights on teaching and learning inclusively. They argue that an education that serves a transformative objective should include the ingredients of democratic participation that necessarily entails dialogic engagement, collective consultation, and community action (Freire 1970; Lewis and James 1995; Nelson and Wright 1995; Maguire 1987). If the ultimate goal of adult education is to prepare citizens for an engaged role in civic life, consultation with the community and their participation is paramount to engendering sustained transformation. New forms of environmental adult education do not remain solely within a framework of a more imaginative training of citizenship which fosters an ethic of collective responsibility and care (Selman 1998), but rather problematizes the ways in which doing so often reinforces social and class inequality. The "circle of certainty" which Freire (1970, 21) cautions against must be questioned so that educators are not blinded by their own understanding of what they perceive to be the reality which informs the process of transformation.

The concepts of 'home' or 'sense of place' expounded by bioregionalists and adult educators are the basis which informs the more inclusive framework of environmental adult education (see Clover 1998; Dodge 1990; Fike 1994, Sanger 1993). They argue that a vibrant community stems from a deep sense of connection to the land or place that leads people to engage actively and conscientiously in the processes of community building. 'Belonging' has been identified as crucial to any movement that aims to transform social or ecological relations (Dodge 1990; Fike 1994). However, processes of imperialism and trans-globality have brought together disparate individuals within post-colonial spaces and reconstituted notions of 'citizenship' and 'home' in their wake. Thus, bioregional thinking must also interrogate the implications of social relations and systems of domination to reveal how they subvert the discourse of community and sense of belonging. Otherwise, bioregional exhortations will remain contained by liberal and utopian sketches of an idealized reality. Many immigrants experience their new 'home' as an unwelcoming space, dominated by people who view them with inferiority and aversion (Tan 1999). The tendency to view immigrants as outsiders undermines their sense of 'belonging' and limits their desire to participate in social processes within their communities.

Critical environmental adult educators resist the temptation of viewing ecological relations as universal norms or monolithic dichotomies of human vs. nature, but rather as contested sites of symbolic and material practices. The 'environment' is a terrain of struggle, a political space weighted by histories of exclusion, subordination, and domination (Rogers 1995). Environmental adult educators are facing the challenge of envisioning approaches to ecological literacy that reflect an understanding and acknowledgement of these dissonances.

Conclusion

The new millennium hastens the reality of the 'global village.' Changing tides of socio-political, material, and ecological conditions are creating a movement in bodies, energies, and ideas around the planet, reconfiguring the social and cultural boundaries of each society and redefining conceptions about the world itself. In Canada, "four out of every five immigrants since 1991 have a first language other than English or French" (Carey 1997), and ethnic vernaculars are slowly taking their place as commonly spoken languages in the country. However, while the rise of immigration has intensified intolerance (Miller 1997), "the denial of racism will not make the problem go away" (Dei 1994, 2). As educators, he argues, "...we should be able to confront and deal more openly with the tensions and uneasiness, the contradictions and paradoxes of discussing race and racism in our [work]."

In spite of the challenges, Canada's cultural mosaic offers much opportunity for re-framing traditional practices of environmental and outdoor education for adults, but we must be prepared to ask difficult questions and challenge our practice and theory. Critical and inclusive concepts of environmental adult education go a long way toward fostering social justice and ecological sustainability for the present and future generations to come. Perhaps its theory and practice can encourage others working within an ecological context to take a greater step toward a movement for eco-social transformation.

References

Agarwal, B. 1992. "The Gender and Environment Debate: Lessons from India." *Feminist Studies 18*, 1, 119–158.

Armstrong, L. 1993. "Connecting the Circles: Race, Gender and Nature." *Canadian Women Studies 13*, 3, 6–16.

Banuri, T., and Marglin, F.A. 1993. "The Environmental Crisis and the Space for Alternatives: India, Finland, and Maine." In T. Banuri and F.A. Marglin (Eds.), *Who Will Save the Forests? Knowledge, Power and Environmental Destruction*. London: Zed Books.

Bodley, J. 1990. *Victims of Progress*, 3rd Edition. Mountain View, CA: Mayfield Publishing Company.

Bookchin, M. 1995. *From Urbanization to Cities: Toward a New Politics of Citizenship*, Revised Edition. London: Cassel.

Bultena, G.L. and Field, D.R. 1978. "Visitors to National Parks: A Test of the Elitism Argument." *Leisure Science 1*, 4, 395–403.

Carey, E. 1997. Mamma Mia! Chinese Beats Out Italians! *The Toronto Star*, December 3, A1.

Chambers, R. 1987. "Sustainable Livelihoods, Environment and Development: Putting Poor Rural People First." Discussion Paper 240. Sussex: Sussex University, Sussex Institute for Development Studies.

Clover, D. 1998. "Adult Education Within an Ecological Context." In S.M. Scott, B. Spencer, and A.M. Thomas (Eds.), *Learning for Life: Canadian Readings in Adult Education*. Toronto: Thompson Educational Publishing, Inc., 213–222.

Cunningham, D. 1998. "White Women with Green Ideas: Research Shows that Godless Caucasian Females are the Strongest Environmentalists." *British Columbia Report, 9,* 18, 25.

Day, B. and Knight, K. 1991. "The Rain Forest in our Backyard." *Essence*, January, 75–77.

Dei, G. 1994. "Anti-Racist Education: Working Across Differences." *Orbit 25*, 2, 1–3.

Dei, G., Hall, Budd L., and Goldin-Rosenberg, D. (Eds.). 2000. *Indigenous Knowledges in Global Contexts: Multiple Readings of Our World*. Toronto: University of Toronto Press.

Dodge, J. 1990. "Living by Life: Some Bioregional Theory and Practice." In V. Andruss, C. Plant, J. Plant, S. Wright, and E. Wright. (Eds.), *Home! A Bioregional Reader*. Philadelphia: New Society Publishers, 5–12.

Elliot, J.L. and Fleras, A. 1990. "Immigration and the Canadian Ethnic Mosaic." In P. Li (Ed.), *Race and Ethnic Relations in Canada*. Toronto: Oxford University Press, 51–76.

Essed, P. 1991. "Understanding Everyday Racism: An Interdisciplinary Theory." *Sage Series on Race and Ethnic Relations, Vol. 2*. California: Sage.

Fike, M.S. 1994. "Food, Farming, Feminism and the Future: An Ecofeminist and Bioregional Vision for Alternative Agriculture." Unpublished MES Major Paper. Toronto: York University.

Freire, P. 1970. *Pedagogy of the Oppressed*. New York: Continuum Publishing Co.

Gibson, B. and Moriah, D. 1980. Meeting the Interpretive Needs of Minorities. *Proceedings of the National Interpreters Workshop*. National Association of Interpreters, St Paul, 223–227.

Gigliotti, L.M. 1990. "Environmental Education: What Went Wrong? What Can Be Done?"

Journal of Environmental Education 14, 2, 19–23.

Goldberg, D. 1993. *Racist Culture, Philosophy and the Politics of Meaning.* Cambridge, MA: Blackwell.

Griffith, D., Pizzini, M., and Johnson, C. 1992. "Injury and Therapy: Proletarianization in Puerto Rico's Fisheries." *American Ethnologist 19*, 53–74.

Ham, S.H., Langseth, R.D., and Fazio, J.R. 1985. "Back to Definitions in Environmental Education: The Case of Inland Northwest Camps." *Journal of Environmental Education 16* 4, 11–15.

Kirby, S. and McKenna, K. 1989. *Experience Research Social Change: Methods from the Margins.* Cambridge: Garamond Press.

Lewis, S. and James, K. 1995. "Whose Voice Sets the Agenda for Environmental Education." *Journal of Environmental Education 26*, 3, 5–12.

Lichterman, P. 1995. "Piecing Together Multicultural Community: Cultural Differences in Community Building Among Grass-Roots Environmentalists." *Social Problems 42*, 4, 513–534.

Livingston, J.A. 1981. *The Fallacy of Wildlife Conservation.* Toronto: McClelland and Stewart.

Livingston, J.A. 1994. *Rogue Primate: An Exploration of Human Domestication.* Toronto: Key Porter Books.

Lousley, C. 1998. *(De)Politicizing the Environment Club: Environmental Discourses and the Culture of Schooling.* Unpublished M.A. Thesis. Toronto: OISE, University of Toronto.

Maguire, P. 1987. *Doing Participatory Research: A Feminist Approach.* Amherst, MA: The Centre for International Education, University of Massachusetts.

Merchant, C. 1992. *Radical Ecology: The Search for a Livable World.* New York: Routledge, Chapman and Hall, Inc.

Miller, A. 1997. Toronto Most Racist City in Canada. *The Toronto Star,* October 6.

Nelson, N. and Wright, S. 1995. "Participation and Power." In N. Nelson and S. Wright (Eds.), *Power and Participatory Development: Theory and Practice.* London: Intermediate Technology Publications, 1–18.

Newman, P. 1994. "Killing Legally with Toxic Wastes." In V. Shiva (Ed.), *Close to Home: Women Reconnect Ecology, Health and Development Worldwide.* Philadelphia, PA, and Gabriola Island, BC: New Society Publishers, 1–19.

Omvedt, G. 1994. "Green Earth, Women's Power, Human Liberation. Women in Peasant Movements in India." In V. Shiva (Ed.), *Close to Home: Women Reconnect Ecology, Health and Development Worldwide.* Philadelphia, PA, and Gabriola Island, BC: New Society Publishers.

Pickering, M. 1995. "New Canadians Enrich Environmental Activism." *Alternatives 21*, 4, 13–14.

Poore, P. 1993. "Enviro-Education: Is it Science, Civics or Propaganda?" *Garbage 5*, 3, 26–30.

Potts, R. 1996. *Humanity's Descent: The Consequences of Ecological Instability.* New York: Avon Books.

Pratt, M.L. 1992. *Imperial Eyes: Travel Writing and Transculturation.* London and New York: Routledge.

Pynn, L. 1995. Vancouver Director Criticized for Racist Comments of Food Fishing. *Vancouver Sun,* September 18, A1.

Ralston, H. 1991. "Race, Class, Gender and Work Experience of South Asian Immigrant Women in Atlantic Canada." *Canadian Ethnic Studies 33*, 2, 129–139.

Rogers, R. 1995. *The Oceans are Emptying: Fish Wars and Sustainability.* Montreal: Black Rose Books.

Seager, J. 1993. *Earth Follies: Coming to Feminist Terms with the Global Environmental Crisis.* New York: Routledge.

Selman, G. 1998. "The Imaginative Training for Citizenship." In S.M. Scott, B. Spencer, and A.M. Thomas (Eds.), *Learning for Life: Canadian Readings in Adult Education,.* Toronto: Thompson Educational Publishing, Inc., 24–34.

Sen, G. 1994. "Women, Poverty and Population: Issues for the Concerned Environmentalist." In W. Harcourt (Ed.), *Feminist Perspectives on Sustainable Development.* London: Zed Books, 215–225.

Shabecoff, P. 1990. Environmental Groups Told They Are Racists in Hiring. *New York Times,* February 1, 11.

Shiva, V. 1994. "Introduction." in V. Shiva (Ed.) *Close to Home: Women Reconnect Ecology, Health and Development Worldwide.* Philadelphia, PA, and Gabriola Island, BC: New Society Publishers, 1–9.

Sly, C. 1991. Letter to the Editor. *Three Circles Centre Newsletter,* 2.

Tan, S.S. 1999. "Tree Hugging and Green Bigotry: Challenging Imperialism in Environmental Approaches." Unpublished M.A. thesis. Toronto: OISE, University of Toronto.

Watson, E.W., Roggenbuck, J.W., and Daigle, J.J. 1992. *Visitor Characteristics and Preferences for Three National Forest Wildernesses in the South.* Research Paper No. INT455, USDA Forest Service Intermountain Research Station.

Wilson, A. 1991. *The Culture of Nature: North American Landscape from Disney to the Exxon Valdez.* Toronto: Between the Lines.

Chapter 2

Re-Invigorating Pastoralist Environmental Practices Through Collective Learning: A Case of Nomadic Ethnic Groups of Northern Kenya

Rosa Muraguri-Mwololo
University of Illinois, USA

The contemporary problems of socio-environmental degradation in the northern semi-desert areas of Kenya are known largely by everyone. One key reason for this growing problem is that the ecological wisdom that has been handed down from generation to generation amongst pastoral people has been systematically eroded and marginalized due to the imposition of 'erratic' colonialism and contemporary economic development initiatives. This means that finding proper solutions to environmental problems is a very complex process. One thing that is becoming quite evident is that socio-environmental solutions will only be found when problems are identified, their causative factors clearly understood, and the people most affected directly and actively involved.

The pastoralist production systems of northern Kenya reflect a complex attempt to balance the co-existence of humans and animals. This highly fluctuating environment in Kenya is characterized by erratic rainfall regimes and uneven pastures and water. While it is now beginning to be more clearly understood that nomadic pastoralism, based on traditional management practices, offered the best land use system for communities, this ecological knowledge has been abused and marginalized for many years.

Many development organizations, including the government of Kenya, are guilty of imposing technologies and innovations on pastoral communities in the name of 'development.' These interventions have heavily eroded and undermined traditional environmental governance practices or systems. Cultural and ecological demands were not met by the new projects and the pastoral economy that is highly fragile was badly damaged. These imposed changes: in addition, they altered the behaviour of the pastoral groups, and traditional strategies relied on in the past for sustainable use of the rangeland resources failed. The result has been degradation of the land around permanent water points and settlement areas and disruption of the previously practised ways of land use that were in harmony with the seasons. The degradation has caused a decline in livestock productivity (the mainstay of this community) and a consequent increase in poverty levels. Heavy losses in terms of human life and animals are experienced during protracted drought periods—an occurrence that was unheard of in the past but very much a part of the present.

To address these problems, a local non-governmental organization (NGO), with the support of the Canadian International Development Agency (CIDA), initiated the Pastoralist Development Project. The goal of this project was to use environmental adult education principles and practice to help to improve the living standards of pastoralists, their capacity to better manage their own well-being and the available resources, and create collective knowledge for change. The strategy adopted involved stimulating discussions with communities about environmental and natural resource management and how it affects their lives, and putting into place locally designed democratic environmental governance structures. These structures are referred to as the Environmental Management Committee (EMC) and the Water Users Association (WUA). Through these committees, the communities are able to regulate the use of available resources by their own rules and regulations based on traditionally tested strategies blended with conventional management approaches.

Greater sustainability and appropriate development for these communities means engaging in community action learning processes, mitigation of negative environmental impacts caused by past inappropriate resource use strategies, and the control and management of resources by those who use them. In addition, policies on land and natural resource ownership in pastoral lands will have to be addressed as a development strategy and as a lasting solution to the emotive issues of resource control and ownership.

Using a case study approach, this chapter analyzes some of the indigenous environmental management strategies previously used by these communities, the inappropriate government interventions and policies that completely disoriented the traditional resource governance systems, and the twelve years (1988–2000) of project experience and educational work used to try to restore the almost lost ecological wisdom of nomadic peoples in Kenya. The approach largely challenges the elitist and foreign models of development, by encouraging and adopting indigenous knowledge and practices in an environmental resource management plan towards sustainability.

Historical and Ecological Background

Northern Kenya is a melting pot of different Cushitic people and cultures, mixing and borrowing from one another for centuries. The inhabitants of the region belong to the same close racial and linguistic stock, described generally as Eastern Cushitic by ethnographers. Their predominant representatives are the Somali- and the Oromo-speaking people, who are in turn divided into smaller units. These groups have often merged into one another, losing and assuming new identities depending on exigencies. Consequently it is sometimes difficult to delineate where one group ends and the other begins. Various Oromo and Somali clans claim ancestry from both groups, while others are bilingual with some sub-sections claiming to be Somali or Rendille. For instance, the Rendille, who are Cushitic, and the Samburu, who are Nilotics, have very similar material cultures.

For a long period of history, northern pastoralists were treated differently from the rest of the people in Kenya. Their pastoral social and political organizations were not seen as conducive to and supportive of the national development strategies. The opinion of some government officials and donors is that pastoral development is too complicated and the returns are too low. This attitude isolated and marginalized the pastoralists, and disregarded the wealth of their knowledge and ways of living on the land.

The ecology of the pastoral dry lands, which is characterized by low, erratic, and unpredictable precipitation, was seen to be simply unviable and technologically infeasible for any kind of profitable development interventions (Umar 1994). The region did not impress colonial administrators as habitable, and the officers were posted to serve in the area on a temporary or short-term basis. Most of those who opted to serve in the region were the adventurous and sporty. A few were induced by a monthly financial reward (the Northern Frontier District Allowance) and other special benefits that were provided for the comfort of those who opted to serve in the remote region. Many efficient development agents were unwilling to serve in the region.

The trend transcended to the independence era, as a legacy, and denied the much-needed services of good officers to enhance the chances of success of the planned developmental interventions for the region. This fear, and the discomfort of the northern climate and environment, was often used as a form of punishment, to post the stubborn, the inefficient, the wrongdoers, which consequently undermined the quality of development implementation.

Nomadic pastoralism is a highly flexible system that has evolved over time as the most effective means of exploiting transient water and grazing resources under ecologically marginal conditions. Nomadic pastoral systems are an ecological requirement for offsetting impacts of frequent droughts and highly erratic and unpredictable rainfall. Climatic conditions in northern Kenya are characterized by high temperature ranging between 30 and 41 degrees Celsius. Annual rainfall averages less than 350 mm.

Contrary to many held notions, pastoralism is neither random nor erratic but based on information generated through traditional surveys and local ecological knowledge of range resources. Pastoral resource use pattern is characterized by risk-spreading and flexible mechanisms such as mobility; large sizes that are diversified, herd separation, and splitting. Pastoralism entails seasonal migration of human and livestock between wet and dry season grazing areas.

Most of northern Kenya is drought-prone. Droughts range in scale and intensity. In many areas, pastoralists distinguish between two types of droughts: droughts that hurt (e.g., cause hunger and shortages) and droughts that kill (cause famine). Pastoralists are vulnerable to droughts and famine, and therefore use a number of indicators to signal when a drought is coming. For instance, elders may read the intestines of slaughtered small stock to forecast changes in weather pattern, wind, clouds, stars, shifts in temperature, and/or changes in an available resource such as well water. They also keenly observe changes in animal behaviour (livestock, birds, and wildlife).

Pastoralist communities in northern Kenya keep a diversity of animals such as goats, sheep, camels, donkeys, and cattle as a drought-coping strategy. Traditionally, camels were taken to *Waiyam* soil after the first rains, to browse until the vegetation dried up. The following year they would be transferred to the *adabla* soils to feed on the nutrition *yumarik (Blepharis linariifora)*. Specific areas are set aside for use by weak animals, those in calf, as well as those in lactation. Other livestock were excluded from use of these preserved areas and any violation was heavily penalized. All animals including camel, cattle, and small stock are divided into wet herd and dry herd. The wet herd, which consists of lactating female stock, calves, and weak and old livestock, graze within a radius of between 20 km for cattle and 35 km for camels. They keep nearer the settlement and watering points. Dry herds move more extensively. Grazing strategies of a given group are largely influenced by their cultural preferences as well as by other factors like schools, markets, medical dispensaries, and at times, the location of relief centres.

The understanding of existing social institutions and resource tenure patterns is often shallow. Development organizations pursue certain development objects in isolation. A fully integrated approach that takes cognizance of all inter-related factors is the only way to sustainably manage in this region. The government of Kenya and outsiders hold strong ideological prejudices against the culture of nomadic pastoralism. A top-down approach to development has been the *modus rivendi* for community needs assessments, participation design, and implementation of programmes.

Many development agents have also held the myth that pastoralism and the pastoral economy are irrational: that pastoralists do not make logical decisions, that they are ignorant about their ecological habits and are only concerned with increasing herd size (cattle complex) for prestige and for spiritual reasons. Another myth is rooted in the western concept of the 'Tragedy of the Commons' that incorrectly stipulates that, since pastoralists tend to compete for available but meagre resources, and livestock are individually owned, they degrade the lands. These so-called personal and selfish interests are believed to supersede those of the common property.

There is yet another misconception. This is the belief that pastoralists do not manage the rangeland but rather use it in a chaotic and random manner. This has led to a very patronizing and paternalistic approach to the planning and implementation of projects, as well as the marginalization of local ecological knowledge.

Government policies since colonial times have not favoured nomadic pastoralist production systems. For instance, during the time that Kenya was under colonial rule, there was an active discouragement of pastoral livestock production in order to protect the European cattle economy. Traditional coping mechanisms used by pastoralists were ignored or given a low priority during planning of development interventions by NGOs and government (Widstrand 1975).

The vast majority of past approaches to arid and semi-arid land development did not adhere to the principle of sustainability. The change in land-use system and settlement of pastoralists led to serious environmental crises. Water development has been a major factor that has greatly influenced land-use change. Schlee (1991), while reviewing the land use strategies of pastoralists of northern Kenya, has recommended the need to encourage pastoralist mobility rather than gradually reducing it through development interventions.

Kenya's pre-independence colonial government used water development and veterinary programmes (Grazing Blocks) to encourage pastoralists to settle. The greatest challenge for the colonial government was to make nomads amenable to governance. The government defined grazing lands and water boundaries. Permanent water sources for livestock use, in form of boreholes, were developed at intervals of approximately 20 km. This resulted in the development of permanent settlement areas and degradation. The Grazing Blocks assumed clan territorial boundaries limited the mobility of nomads whose production system is based on flexible resource utilization. The heavy concentration of pastoral numbers, rather than alleviating environmental degradation, had the reverse effect. The 1936 drought exposed the strains of colonial policies that discouraged home-grown coping strategies. The colonial government's development policies were not only piecemeal, but also lacked any form of good will.

Development motives of the colonial government are still a subject of ridicule but the irony is that post-independent Kenya (1963) inherited and continued to practise these colonial policies with little or no change. Most of the adopted policies flew in the face of reality in general, and indigenous ecological knowledge and experience in particular. According to Abound (1986) rangeland management projects funded by external agencies were not at all successful. Settling pastoralists simplified administration, but this confinement of herds in ecologically unviable land units caused extensive environmental damage and a serious loss in biodiversity, particularly in the country's north. Undue pressure placed on one point or area tended to reverberate across the entire ecosystem. In some instances, intense overgrazing gravely affected soil conditions, texture and moisture content, and subsequently led to the loss of plant species. As vegetative cover thinned out, access to wood sources for fuel also diminished. This caused increased competition for resources and invariably momentum for further degradation and species loss was accelerated. So-called desert towns emerged and are mainly composed of those who are unable to survive in the pastoralist system and are condemned to a pathetic lifestyle of petty trading in *qat*, or hopelessly waiting for relief food.

Misinformed and insensitive development policies have accelerated environmental destruction in the arid and semi-arid lands (ASALS), resulting in some cases in major ecological changes. As mentioned previously, the assumption that pastoralism is inherently a destructive land management practice, that pastoralists in fact represent a liability to the national economy and are therefore in need of external intervention, has been disastrous. It is as a result of these changed perceptions that new alternatives and approaches to development, based on the past and people's ecological knowledge, are emerging.

Slowly and painfully, development partners are beginning to agree on one thing: there is a need to restore traditional practices in 'pastoral democracy' if there is to be a sustainable use of range resources in Kenya. Moreover, for better environmental protection, there is a need for formal acknowledgement and respect of tribal lands, and tribal people's ecological knowledge and ways of living with the land. In addition, the customary good relations between tribes must be supported.

To begin to solve the problem of inappropriate resource use, the Pastoralist Development Project, with the support of CIDA, has encouraged and facilitated the formation of community-based institutions for the control and management of the natural resource base. These institutions are actually structured on the experiences and ways of working of traditional pastoralist institutions or practices. The institutions supervise, direct, and control resources, and provide advice on strategic development activities. Each community develops its own guidelines and regulations regarding the use of resources in their locality.

This project pioneered new strategies in environment and natural resource management, specifically, community-based management of grazing resources. Resources were allocated to train pastoralists on how best to allocate and manage resources in a sustainable manner. The first phase of the project focussed on range resources inventory and gender-analysis—an approach that was used to determine the status of the forage and general trends over the years. The information collected was used to develop strategies for community capacity building.

Learning and Education in Resources Management
Approaches for Pastoralists

Since pastoral land is owned and used communally, the most effective approach to resource management was to work through groups to reach communities. Using highly participatory bottom-up methods, the groups work out their common problems and solutions as a team. Owning the process, from active participation at the planning level through to the implementation stage, the communities create and thereby identify with the management systems or practices adopted.

Agenda 21 (UNCED 1992, 20) states that environmental problems will best be handled through the full participation of all concerned citizens, and that women have a vital role in environmental management and development. Based on this, an inclusive approach was adopted that takes the concerns and roles of the various categories in the community structure, that is, the elders, the women, the men, and the youth, into account. In particular, the active participation of women in the design and implementation of the programme was seen as a novel idea in this deeply patriarchal community.

Past learning and education practices in many communities in Kenya were often done through stories, songs, and the sharing of rituals. The British colonizers condemned these as barbaric, savage, and primitive. It is little wonder that many Africans have lost their closer relationship with the rest of nature from which they used to draw support for survival. With the erosion and marginalization of learning with, about, through, and on the land, of indigenous ways of knowing and its code of conservation, the respect for and appreciation for the rest of nature for its own sake and in its own right, is nearly all but lost.

The training or learning process being introduced involves going back to that knowledge and weaving in new ideas to create new knowledge. It is based in discussing, with the community, the hazards of land degradation as they themselves experience it firsthand. Discussion focusses on a range of problems, such as loss of soils through water and wind erosion, loss of vegetative cover due to overgrazing, decreased availability of water, and a general reduction in the diversity of plant life. People were invited to share their views and experiences in terms of how these factors increase their vulnerability to drought and food security.

The learning programmes entail visits to target groups, having community days, and visioning workshops where the facilitator guides the participants to reflect on past wise use of resources, current use, and likely future trends. Through this learning process, participants were able to more fully envision and reflect upon the complex problems they face and, from there, develop possible courses of action towards an integrated, comprehensive, and sustainable approach to the management and use of resources. Sustainable ways of working with resources can address both environmental and development needs. In addition to this activity, methods of negotiation and conflict resolution techniques are thoroughly discussed. Leadership, accountability, responsibility, and principles of good governance are important aspects of the overall training strategy. The key emphasis throughout is to encourage resource users to rely on past traditionally tested strategies whilst incorporating conventional modifications, where and when appropriate.

The next important stage is to form a community-based management or governing structure which can supervise, direct, and monitor the use of resources in a just and equitable manner. Using highly democratic and transparent methods, community members elect a group of people to act as community custodians. As a safeguard to ensure that the process truly is openly democratic and acceptable to all in the community, local leaders such as politicians, chiefs, assistant chiefs, and opinion leaders are invited to be present.

Once the elected members take office, they are required to organize community forums to create the rules and regulations to safeguard the use of resources in a sustainable manner. The facilitators take a back seat during the election process but assist the community in the streamlining of rules and regulations where necessary.

To ensure that there is consistency in approaches, all development agencies including government departments are involved in the process. Through consensus, the committees were named the Environmental Management Committees (EMC) and the Water Users Associations (WUA). Women, as major resource users, are highly represented in the both committees. The total number of EMCs and WUAs in these districts is increasing as the other NGOs have also helped in setting up the same with other communities.

Roles of Environmental Management Committees (EMCs)

The EMCs have a number of responsibilities. First, they are the entry points to the community for any development agents on matters relating to environment and natural resources. Development agents can use them to mobilize people for training or discussion with the community but must respect the ways in which the community works. Second, the EMCs must engage in the formulation of rules and regulations and ensure that they are not abused by the community. They are responsible for collecting penalty fines from defaulters and user fees for resources they charge for, for example, construction poles for houses. The EMCs also define areas of grazing, time of use, and adjustments to grazing patterns within seasons for the community they represent. Fourth, the EMCs identify and assess development needs and strategies for implementing, supervising, monitoring, and evaluating any community projects relating to grazing resources and water developments. This includes negotiating with neighbouring communities on behalf of the communities for resource sharing at critical times; resolving any conflicts which arise relating to grazing and water; and regulating the influx of neighbouring livestock to areas of their control.

The EMCs are responsible for making links with WUAs to ensure that the water points are opened and closed in harmony with the season and adopt a grazing strategy. The EMCs also control the activities of WUAs and have the final word on who is to use a certain water point, especially the boreholes.

Roles of Water Users Associations (WUAs)

The WUAs have other types of responsibilities. One is to ensure that the use and management of specific water points is appropriate. They operate under the supervision of EMCs. Allocation of livestock watering time for all the livestock owners that are allowed by the EMCs to use a certain water point also falls within the mandate of the WUAs. The WUAs engage in the collection, recording, and banking of the user fees they collect. They ensure that the water points are cleaned after use, and that there is regular servicing and maintenance of boreholes.

Generally, by setting these community-based management structures, there is increased sense of responsibility and accountability in the use of resources. Where the committees are operating effectively, there has been a shift from an 'open access' use of resources to 'communal use,' thereby ensuring better and more sustainable usage.

Experiences with Community-Based Organizations (EMCs and WUAs): Lessons of Experience

After working with community-based organizations in twenty-four communities for a period of about four years, the following positive changes has been observed:

- Communities now have a greater capacity to control and improve efficiency in their use of resources. Reverting back to more traditional mechanisms or systems and linking these to new approaches where appropriate has proven to be highly affective.
- Once the communities identify with the resources base in their locality, any intervention aimed at improving the use and undertaken with the full participation of the target community is unlikely to generate conflicts. The use of available grazing resources and adherence to seasons of use has been encouraging.
- Community-set rules and regulations are readily adopted and respected by the resources users. In cases where there are defaulters, it has been observed that the person who has offended the other users pays the penalty fines set, and in this way, similar incidents are discouraged.
- As a result of the existence of EMCs in most communities, and the general awareness among pastoralists of the existence of such institutions, there has been a general decline in conflicts over grazing rights, especially within areas where different communities often compete for resources. However, tribal animosity aggravated by political interference has at times interfered with peace initiatives.
- Once community institutions are established, it becomes easy to mobilize target communities to undertake other development projects with little or no external support. Cost-sharing policies with development partners has increased with the new knowledge. Paternalistic and patronizing, top-down approaches to development are now being questioned by communities. The new approach has been highly empowering!

Any future policy for developing rangelands of northern Kenya must recognize and take into consideration the ecological and socio-economic forces that underlie the survival strategies of pastoralists who utilize them. The arid pastoral ecosystems of northeastern Kenya should be managed for their adaptability rather than stability. In brief, there is an absolute need for formal acknowledgement and respect of tribal lands, and the customary good relations of tribes to be supported. This has and will continue to greatly reduce inter-clan conflicts over resources. 'Pastoral democracies' should be encouraged to foresee sustainable use of range resources.

This type of learning and development initiative, to draw out traditional ecological knowledge and build community capacity, is relatively new in northern Kenya. Pastoralists move a great deal and many development agencies found this 'behaviour' incompatible with their rigid approaches to development. The Pastoral Development Project has managed to overcome this problem by simply adopting new ways of following the communities. The better uses of resources in this project provide evidence of an increase in livestock production, a decrease in resource-related conflicts, and better preparedness for times of drought.

Constraints on Community-Based Approaches

Development in the arid and semi-arid lands occurs in an environment characterized by many known constraints. In northern Kenya, major constraints include:

- Characteristically low rainfall regimes that limit forage production. The situation is aggravated by frequent rainfall failures, unpredictable weather patterns, and droughts. Under such environments, it is at times difficult to strictly follow the communal management schedules in relation to places and time to be spent in a particular grazing site. At times, the animals are moved to the only sites that have received rains regardless of season; during drought situations, survival for both livestock and human beings becomes the main criteria for influencing the decisions taken by the livestock owners.
- Due to the expansiveness of the area, decisions of the EMC may be limited to only areas around the settlement vicinity. In some situations, areas reserved by one community as dry season reserve may be used by another group in the wet season without the knowledge of the EMC. At times, conflicts may arise on strategic grazing areas and reserves as some communities are more influential than others. Also, in instances where households require baggage animals (camels and donkeys) to move regularly and over long distances, families without adequate transport find it difficult to keep up with those with adequate means. This poses difficulties for the affected families, and they may find themselves in conflict with EMCs, who are mandated to see that every community member observes the set rules and regulations.

- Cultural practices such as circumcision ceremonies that require the indiscriminate cutting of trees can sometimes undermine resource management and destroy the environment. The complete destruction of trees around sites where such practices are performed has been observed in Samburu and Rendille communities.
- Insecurity in pastoralists areas has become a major constraint to pastoralists' production system improvements. Pastoralists avoid large areas in northern Kenya due to fear of attack by hostile or neighbouring communities. This usually has an effect on the management systems and movement patterns of pastoralists, thereby reducing the efficiency in use of available resources. The political environment in which the issues of insecurity are handled has continued to complicate and aggravate the situation. Any gain in production improvement is greatly decimated when some communities lose their lives and livestock due to raids.
- There is generally poor infrastructure development in most arid and semi-arid land (ASAL) areas, thereby making provision of essential services difficult to remotely located pastoralists. This tends to encourage pastoralists to locate their household bases near developed centres so that they can be near shops, schools, and dispensaries. They also camp near centres to access famine relief, which they have come to include as part of their requirements.
- The fact that nomadic pastoralists are relatively unpredictable in their movements and wonder about in search of resources poses one of the greatest challenges to development workers. However, the Pastoralist Development Project was able to surmount that challenge through the creation of mobile outreach services and mobile outreach camps. By following the nomads, they were better able to bond and ensure the continuity of the environmental adult education work.

Recommendations

At the risk of sounding repetitive, it is important to not only highlight the positive changes and challenges, but to assemble them into specific recommendations which may benefit other development agencies who wish to use this form of environmental adult education in their own work in Africa.

For environmental adult education to be most effective, the learning process must be engaging, participatory, based in people's local knowledge, and critical. The learned must be engaged not just cognitively, but through their imagination, senses, passions, and feelings. This means that people's everyday experiences and their wisdom and knowledge must be the fundamental building block of the learning process to collectively create new knowledge for change.

The most significant step toward pastoralist development is to ensure that resources are managed and controlled by the resource users. This would entail defining user groups in all areas, such that all grazing areas are identified as belonging to a particular community or tribe. Neighbouring communities and tribes will hence have to negotiate with the defined owners to share resources. With the support of local community leaders and administrators, development agencies, and politicians, such a move would reduce or minimize hostility between groups and tribes, as resources would be identified with owners. Caution must be observed when allocating resources, as there is a tendency for some communities to stake out what does not belong to them traditionally.

To cushion the impact of droughts and adverse weather, there is need to facilitate pastoralist communities to diversify their economic base. Opportunities existing within the local environments should be exploited. Alternative food production systems (e.g., small scale irrigation projects, crop production using water harvesting techniques, and agro-pastoralism) should be tried where feasible. This should also be done with a great deal of care to ensure that they do not affect the sustainability of livestock production. Other sources of income for pastoralists like bee keeping, trade in artifacts, eco-tourism, ostrich farming, micro-enterprises, and other income-generating activities should be encouraged in pastoralist areas.

The traditional coping mechanisms used by pastoralists were unknown, ignored, or given a low priority during planning of development interventions. There needs to be ongoing recognition of this ecological knowledge and wisdom, and learning processes put in place which draw upon and strengthen it.

Despite there being information on ASAL areas, there is still an information gap in critical areas of pastoralist development. Little is documented on the interface between traditional knowledge and conventional or technical knowledge and how experiences in other ASAL areas of the world can be applied to solve existing problems. A thorough study is required on appropriate resource management systems for ASAL areas in light of the damages that have continued to occur in these areas. Although there have been claims that the pastoralist production systems are understood, it is doubtful whether the application of such knowledge is as straightforward to development agents. There is a need to conduct more research in areas of traditional resources management, traditional ecological knowledge systems, human livestock rations, social changes, and impacts of development on ASAL inhabitants. Moreover, future research should be conducted with farmer or herder participation to facilitate understanding and implementation of results.

ASAL development projects should conduct proper environmental impact assessments before they are implemented. This should be encouraged because it would help in updating the inventory of range resources, and incorporate mitigation measures of negative impacts of the project.

There is a need to review the present land policy as it relates to pastoralist land use. The current land policy (state land whose power over control is vested to the local authorities) does not give inhabitants of arid lands legal basis for proper control of resources. This has encouraged misuse. Presently, communal land ownership remains the most viable option for control of pastoralist grazing lands. The formation of EMCs is seen as a step in the right direction.

Vigourous education programmes should be conducted in pastoralist areas on environmental issues and implications of pastoralist activities to the environment, the economy, and global environmental problems. Environmental adult education creates avenues for alternative sources of livelihood, and also enhances and harmonizes transfer of information and technology between pastoralists, researchers, and extension/development agents.

If meaningful development is to be achieved in pastoralist areas, there is a need for coordination and consultation between various institutions involved in development of ASAL areas. This would provide an opportunity for institutions to pool resources to avoid or minimize duplication of development efforts for the benefit of the target beneficiary communities.

To conclude this section, it is important to raise the issue of literacy. Although a lot of best practices are documented and the Pastoralist Development Project has produced 'how to' manuals, nomadic pastoralists in Kenya have the lowest rates of literacy. Greater sustainability will best be realized when such issues are seriously addressed. A more literate and numerate community is more likely to understand and follow facts, giving them more confidence through the ability to articulate issues important to their own lives and development. Greater access to information will enable individuals to better understand and to participate more effectively in the decision-making processes in their communities and countries. Information is perhaps the most powerful tool available to people, one that opens up new possibilities for the exercise of both rights and responsibilities. Basic education increases people's capacity to learn and to interpret information.

However, the relevance and value of the formal education given to the nomadic pastoralists have been questioned severely. It is deemed as not only culturally insensitive but also inappropriate. It is important to explore alternative approaches to non-formal education which are flexible, participatory, and perhaps most importantly, culturally relevant. It is important to expand opportunities for functional literacy for adults and to critically address external socio-environmental issues that affect participation in education. Most importantly, it is critical to look at different approaches or strategies to adult literacy practice in Africa, particularly those which revolve around the concept of 'ecological literacy,' of learning to read and write with, about, through, and for the land (for an example, see LEAP 1998).

Conclusion

Environmental problems in Kenya are closely related to the misuse and mismanagement of natural resources and ecosystems. Many of the environmental problems Kenyans face today are as a result of colonialism or the imposition of western social values in the form of education and development, and the insidious adoption of these within the local systems. These have undermined traditional African environmental management practices and the wisdom that supported them, both of which suited local conditions (Lamprey and Yussuf 1981).

Today, the vulnerability of the natural environment is compounded by the fact that so much of Kenya's land mass is arid and semi-arid, and therefore prone to drought and desertification. As mentioned previously, knowledge about the care of the environment in many communities in Kenya was passed on to new generations through stories, songs, and rituals that were all condemned, eroded, and marginalized with the coming of the British colonizers. It is little wonder that many Africans have lost their close relationship with the rest of nature from which they drew support for survival. The appreciation for nature for its own sake and in its own right is nearly all but lost. The erosion and marginalization of indigenous knowledge, wisdom, and more appropriate codes of conservation have led to an increased exploitation of the rest of nature.

The precedent has been that most educational planners have disregarded people's ecological knowledge and this has led to the development of programmes that emphasize 'awareness raising' which totally ignore the fact that people have a wealth of knowledge that has come from a variety of sources, including daily experience. These programmes often fail to foster better practices in resource management primarily because they are based on the assumption that 'awareness' is missing, and it is the only factor related to people's behaviour toward the rest of nature.

The Pastoralist Development Project attempts to tap into people's existing knowledge and, by using experience and traditional practices, actively engage pastoralists in a process of learning and education that is engaging, participatory, and based in their home place, no matter where they travel.

Observations of this learning approach with twenty-four communities demonstrate that, although not without its problems, a major improvement in environmental resources management is coming about. Greater control and management of the natural resource base has reverted to specific communities in each locality and this has enhanced a sense of communal ownership and accountability. Inappropriate practices in resource use have thus been minimized. This implies that there is a need to support these types of non-formal environmental adult education efforts in Kenya.

Note

The author gratefully acknowledges the input of Wamugi, an environmental officer with the Pastoralist Development Project.

References

Abound, A.A. 1986. "Range Development Constraints and Research in Kenya." In R.M. Hansen, B.M. Woie, and R.D. Child (Eds.), *Rangeland Development and Research in Kenya*. Proceedings of a conference held at Agricultural Resources Centre, Egerton College, Njoro, Kenya, April, 365–372.

Lamprey, H.F. and Yussuf, H. 1981. "Pastoralism and Desert Encroachment in Northern Kenya." *Ambio 10*, 131, 134.

LEAP 1998. "Ecological Literacy in Africa." *Pachamama 1 and 2*.

Schlee, G. 1991. "Traditional Pastoralists—Land Use Strategies." In H.J. Schwartz, S. Shaabani, and D. Walthers (Eds.), *Range Management Handbook of Kenya* II, 1. Marsabit District, Ministry of Livestock Development, Nairobi, 130–164.

Widstrand, C.G. 1975. "The Rationale of the Nomad Economy." *Ambio 4*, 4, 146–153.

United Nations Conference on Environment and Development (UNCED). 1992. "Agenda 21." Rio de Janeiro.

Umar, Abdi. 1994. "Sustainable Development for North Eastern Kenya." November.

Chapter 3

Indigenous Struggles for Forests, Land, and Cultural Identity in India: Environmental Popular Education and the Democratization of Power

Dip Kapoor
McGill University, Canada

Introduction

The Kondh *adivasis,* meaning 'original settlers,' of the east coast state of Orissa in India have been struggling to secure cultural and physical space for themselves as they are increasingly marginalized by the process of national development and modernization. This struggle found a new footing in 1995 when thirty-five *adivasi* villages formed a partnership with a small local *adivasi* non-governmental organization (NGO) established by *adivasi* activists from the region. Five years later, this partnership has grown to include some 8000 *adivasis* located in 60 villages. The activist-educators from the local NGO have consciously employed environmental popular education to facilitate a process of social change that aims to re-affirm indigenous cultural identity and promote community control over land and forests that are vital for the cultural and material survival of the Kondhs of the region.

This chapter introduces the social activism of an indigenous group engaged in collective action and environmental popular education that tests the constraints and barriers commonly encountered during processes that attempt to democratize entrenched configurations of power (Freire 1994). It analyses how environmental popular education is being utilized by *adivasis* to encourage them to assert themselves to protect their way of life, to challenge and reconfigure power in *adivasi*-state relations, and to protect the environment. Particular attention is directed to an analysis of the prospects for reconfiguring entrenched power and preventing ecological destruction through environmental popular education and *adivasi* struggle in small, localized movements. Critical reflection on this experience addresses theoretical and methodological issues concerning environmental popular education and prospects for the democratization of power, and the protection of the environment.

The Context

The state of Orissa lies along the eastern seaboard of India, by the Bay of Bengal. Mining, fisheries, agriculture, and forest-based industry account for the primary economic activities in the state. Nearly 90 percent of the state's 33.5 million people live in villages, and close to a quarter of the population belong to one of the sixty-two different tribal or *adivasi* (original dwellers) groups. The Kondhs are the largest of these groups with a population of over 1.25 million people (Nayak et al. 1990). As is the case with indigenous people around the globe, the Kondhs struggle to protect their forests and their *aranya sanskriti* (forest culture) in the face of continuous pressure to alienate from the forests, in the interests of 'development' projects that disregard natural and social limits.

The Kondhs inhabit most of Phulbani district and extend from there into the neighbouring districts of Gajapati, Kalahandi, Balangir, Dhenkanal, Rayagada, and Koraput. Kondh villages are located in the hilly eastern ghat range and on the flat valleys of the ghats at 2000–4000 feet above sea level. According to sketchy historical and anthropological records, the Kondhs retreated into the inhospitable hill-tracts and forests of the ghats from the fertile plains of the Mahanadi River to escape, first from Dravidian and then Aryan invaders, at various times since 4000 BC. This retreat was accelerated by development during the British colonial period and after India achieved its independence in 1947 (Nayak et al. 1990).

As primarily a hunter-gatherer society, Kondhs have always relied on the forests. However, this relationship and their way of life is being increasingly threatened by deforestation and a discriminatory land classification scheme that has its roots in the British Forest Act of 1865. This act declared the forests to be the property of the government and established the basis for post-independence state forestry policies that have promoted the commodification of the forest as a source of revenue and profit for the state and industrialists, who have a common interest in creating a modern and developed independent India. Historical and traditional local-use rights to the forest for subsistence by India's 427 constitutionally recognized 'scheduled tribes' (*adivasis*) has been undermined by a complex system of land classification that has essentially led to the detribalization of tribal forests/land. *Adivasi* rights to the forest have been reduced to grudging concessions or outright denial of tenure or access (Guha and Gadgil 1989). People for whom the forest is a place of spiritual and cultural significance, and the basis for a sustainable moral economy of provision for subsistence, have been reduced to the status of 'encroachers' in order to make way for the unsustainable development agenda of the Indian state.

According to Guha and Gadgil (1989), the removal of forests from the traditional *adivasi* moral economy of provision and their insertion into the modern state-industry political economy of profit has undermined the conditions conducive to sustainable use of land and forests. For instance, while 17 percent of India's land area was forest in 1972, by 1990 only 10 percent of total land area remained forest. Four million hectares of forest were 'lost' between 1951 and 1976, while 23 million hectares were cut down in the next 15 years (Gadgil and Guha 1992, 196; UNDP 1992, 173). While forests covered 45 percent of Orissa in 1971, the figure had declined to 35 percent by 1994 (Oxfam 1996). Such large-scale clear felling has made even limited local use of the forest ecologically hazardous (Baviskar 1995). The *adivasis* are victims of state laws that compel them towards desperate unsustainable forestry practices. Yet, while they are blamed for environmental degradation that results from these practices, the much greater impact of state/market-related forestry is excused as 'development.'

State control over forests has led to pressure on the area of land available for cultivation. *Adivasis* are increasingly taking to agriculture to make up for loss of provision from the forests, but their rights to land tenure are also problematic. For example, according to one study, 200,000 *adivasi* families have no rights to the land that they cultivate (Chambers et al. 1989). Such policies have led to cultural alienation, impoverishment, and increasing dependence. Nevertheless, many *adivasis* still struggle to remain independent, self-provisioning, and agents of their own histories.

Environmental Popular Education

Socio-cultural theories about indigenous groups and the environment accord *adivasis* or 'subsistence producers' the privileged status of belonging to societies based upon a 'life-enhancing paradigm.' *Adivasi* culture has survived over centuries because they have learned to "be like the forest, sustaining both the forest and the culture through time." Their lives and 'traditional wisdom' embody an environmental consciousness and a 'critique of development,' as they live in and with nature (Shiva and Bandyopadhyay 1990, 77; Shiva 1989, 47).

Critical theorists in adult education point to the significance of new social movements as new learning sites, where local autonomy and environmental movements engage in a praxis that challenges and democratizes power. They argue that power has become concentrated in the 'system world'—a world that increasingly dominates what Habermas (1984; 1987) refers to as the 'life-world.' Within this view, new social movements are seen as protagonists, as 'new' agents of history and social change, who are concerned with a localized cultural politics. They are defenders of the life-world as opposed to agents who aim to take over the state (as in the Leninist dream) or overthrow the existing relations of production (as in the Marxist dream) (Welton 1993).

These theoretical assumptions about indigenous peoples, their agency in the social change process, their culture/wisdom, and their environmental consciousness help to inform environmental popular education practices. As such, they emphasize the importance of a holistic view of nature, politics, society, and culture in the concept of 'environment' and educational approaches that emphasize learning through experience and all the senses (Clover 1995). Such approaches stand in contrast with the rationalism of critical andragogy that some critics claim can produce a 'monoculture' of ideas (Shiva 1993).

Popular education is based on the premise that the act of teaching and learning is political, and that the purpose of a dialogical problem-posing education grounded in people's experience and local knowledge is to democratize and reconfigure concentrations of power that are oppressive and dehumanizing (Hall and Sullivan 1994). Educators and learners come together to engage in critical reflection to analyze contradictions in local experience and promote heightened consciousness of the role of structural explanations for the oppression that dominates individual/local experience. This creates the possibility of praxis in the form of collective responses aimed at transforming oppressive power and asserting local aspirations. Popular education aims at the "unveiling," the revelation of "objective reality" (Freire 1970) and creates the basis for a "new apprehension of the world" and a "disposition to change it" (Freire 1994, 145). However, such *conscientization* is authentic only when it "constitutes a dynamic and dialectical unity with the practice of transformation of reality" (103).

The deconstruction and transformation of social reality requires engaging in a critical analysis of power because social life is a product of power (Foucault 1980). Power is located in rules and regulations (Giddens 1984) as well as in the economic and political structures that dominate human communication (Habermas 1984). Power operates in discourses and practices and is not "localized here or there, never in anybody's hands, never appropriated as a commodity or piece of wealth" (Foucault 1980, 98). Power can never be eliminated for, as Foucault argues, what is regarded as truth is constituted through power. However, it can be subverted and transgressed in the interest of disempowered groups. This is a defining element of popular emancipatory praxis (Inglis 1997). Alternatively, for those who deem it plausible, understanding how power works is crucial for reaching truth by overcoming self-interest and the domination of economic and political power in our lives, and in developing a society based on free, undistorted communication (Habermas 1984; 1987).

Being able to transform social life necessitates being able to understand different types of power and the ways in which they operate in society as a whole, as well as in the lives of individuals. The only freedom or emancipation comes from resistance and turning power back on itself. Popular education and the unveiling of power through critical dialogue not only offers a powerful method, but also provides an opportunity for placing environmental popular education and environmental discourse in a socio-political context that links the practice of power to social relations, and to human relationships with the rest of nature.

Environmental Popular Education and Knodh *Adivasi* Assertion

A group of ten (four females and six males) *adivasi* cultural workers, or activist-educators as they will be referred to in this chapter, formed a local non-governmental organization (NGO) in 1995 to work with their people to support the *adivasi* way of life, for which the forest and the land are of primary significance. What started as a partnership with a few Kondh *adivasi* villages soon grew to an organized group of 60 villages and some 8000 people in the region. Conscious of the link between the 'natural' and 'social' crisis experienced by *adivasis*, these activist-educators grounded the process of environmental popular education in the synergy between deforestation and land/cultivation on the one hand, and questions of disempowerment, silence, cultural identity, and loss of the means to reproduce a material existence, on the other. In this way, environmental popular education became a key strategy in the process of *adivasi* articulation and assertion.

The emergent themes from this environmental popular education process are based on field notes written by these activist-educators and the author (as participant observer) as they engaged with an increasing number of Kondh villages. The focus here is on the forest- and land-based concerns expressed by the *adivasis* and the activist-educators, even though dialogical problem-posing education often travelled down other paths during these exchanges.

Traditional Codes and Metaphysical Explanations
for Deforestation and Landlessness

The dialogical problem-posing process of environmental popular education involved discussions with Kondh villagers about the everyday joys and hardships of being an *adivasi*. These exchanges about shared experiences helped to identify generative themes and associated codes related to deforestation-aeforestation, *podu* (traditional shifting cultivation), landlessness, and hunting and gathering, as opposed to cultivating. These themes are codified into 'traditional songs' developed by the activist-educators, and in ritual symbols such as the *meriah* post (for animal sacrifices to the earth deity), and the *Darni* stones (symbol of *Darni*, the earth deity) located in each village. These 'traditional codes' are used to stimulate dialogue, problem identification, and an analysis of the contradictions in the lived experience of *adivasis* as it pertains to their deteriorating relationship with the forest and the land and the gods and ancestors who are the protectors and diviners of their well-being/destruction.

In decodifying 'traditional codes' such as *Kui* songs (Kondh language), the *meriah* post, and the *Darni* stones, activist-educators are drawing on the spiritual beliefs of the Kondhs to explain why the forests are disappearing, and why land and *podu* are becoming problematic. Such activities draw on the Kondh belief that *Darni* came out of the Earth as the first Kondh woman and then, at her own request, became the first human *meriah* (sacrifice) in order to ensure fertility of the forests and land for the Kondh people. *Darni* holds the dual role of deity and symbolic founder of the Kondh people.

Lineal descent from her and from the land gives Kondhs the conviction that the forest and the land are the natural source of all life/their life, and their reason for living. As a result, the gods, the spirits of ancestors, all living creatures, the natural environment, and the past and present are united in a sacred trust for the Kondhs, a trust that is justified by the central significance of *Darni*, the earth deity and creator.

Activist-educators point out that Kondh forest and land use patterns were shaped by these strong spiritual beliefs. In earlier times, Kondhs were strictly forest dwellers and hunter-gatherers. The hills and forests around a village were a common responsibility and, even after cultivation began to be a way of life, a site with uncut trees has always been dedicated to the village deity. Forests cleared for cultivation were allowed to regenerate through a pattern of shifting cultivation; after-use cultivated land was left untouched for ten years or more to allow the forest to regenerate. No fruit or shade trees were cut, and roots were left in the ground to conserve the soil, and later regrowth where trees had been cut down. Environmental popular education reinforces the importance of these traditional approaches to using the forests and the land.

Adivasis also discuss the life-sustaining virtues of the forest. The forest, they remind each other, provides most of their herbal remedies (from bark, roots, leaves, flowers, and fruits), housing materials (timber, bamboo, grass, and creepers for rope and thatching), and food (edible roots, mushrooms, leaves, shoots, flowers, fruit, and honey) as well as tough leaves for plates and bowls, strong reeds for mats, and split bamboo for plaiting cots. Fibres from certain trees and bark are also processed to make clothes. Such discussions emphasize the symbiotic relationships between the forests and Kondh life.

The activist-educators often introduce a *Kui* song about felling trees for cultivation to shift the discussion from the forest to cultivation, *podu*, dispossession and landlessness. Kondhs often express a sense of 'guilt' about the shift from a hunter-gatherer subsistence economy to *podu*. The priest reminds participants of the significance of the *meriah* sacrifice to the earth deity in this regard, reciting the opening lines of the ritual: "We cultivated your Earth, we defiled you, so take this buffalo and eat it! Make our corn grow, don't bring sickness and trouble, et cetera."

Deforestation, landlessness, dispossession, and crop failure are explained in terms of failing to be worthy of *Darni's* protection, and reaping the displeasure of ancestors, because the Kondhs have failed to maintain the right balance between their lives and their spiritual commitment to the natural world. Action to alleviate this situation would demand renewing these connections through ritual and sacrifice, and through greater efforts to seek the grace and protection of the ancestors. Activist-educators appeal to the 'power of *Darni,*' and through her, 'Kondh power,' to regain their rightful place in the forest and on the land that is hers and theirs to protect, rebuild, and conserve. "What are the obstacles?" they ask. "And why have we lost the power to protect and preserve the forests and our way of life? Why do we feel powerless to do anything other than to cut down the forest (our home)? Why do we migrate to look for work to sustain ourselves and our communities? What is compelling us to destroy and leave our way of life?"

Contemporary Codes and Material Explanations
for Deforestation and Landlessness

To focus on contemporary dilemmas, activist-educators introduce a code such as the receipt for a fine for *nevad* (literally means 'new land') cultivation on land 'owned by the state.' The receipt is often on the back of a cigarette box so that the revenue officers can 'pocket' the fines. In discussing such experiences, *adivasis* begin to express their incredulity at the suggestion that they can encroach on land from whence they were born! They discuss why they should be punished for working on land that their ancestors cultivated with their blood and sweat.

The Kondhs tell stories about their experiences with agents of the Forest Department and the Revenue Department and how, in some cases, their small fields and crops of millet and legumes were cleared for teak plantings without any warning from the state agents. They speak of paying 'fines' in money or in kind (chickens) to revenue inspectors who promised to look the other way in return for these payments but who returned regularly to extort more for the same 'crime.' They often speak of a feeling of powerlessness to deal with these 'powerful outsiders' who are literate and able to back up their accusations with alien legal stipulations and the very real ability to arrest, cuff, beat, and jail those that do not acquiesce to their demands. They often express their frustration by saying they would like to beat up these officers, but also say that they are speechless when they are face-to-face with one of them.

Activist-educators begin to question the frustration expressed in talk about beating up officers by asking the group to consider why officers can get away with this kind of treatment. Where do they get the power to clear land being cultivated for years by the Kondhs? The response usually relates to the government's ability to enforce its laws, the relative powerlessness of *adivasis* to keep up with fines, harassment, beatings, crop burnings, and confinement, and their inability to understand the laws in whose name such actions are justified. Such lines of inquiry provide a review of the history of dispossession and the role of the state and the law. The activist-educators remind the *adivasis* of their history of dispossession by the early invaders, the British, and then the Oriya/Hindu 'outsiders,' and how they have gradually been pushed from the plains into the more remote forests and hilltops, while the invaders secured the best tracts of land for themselves. They discuss how laws were framed and implemented by the colonial government and then the government of India. They explain the role of the British Forest Act of 1865 (that declared the forests to be the property of the government) and post-independence amendments to the Act. The end result of such laws, they explain, has amounted to the expropriation of the centuries-old rights of the *adivasis* over the forests, and the detribalization of tribal land. And all this, they point out, has been accomplished in the name of tribal development projects and for the welfare of *adivasis*.

They continue to explain that 'village forests' have been established to provide *adivasis* with timber for firewood, land for grazing cattle, and for other local needs—and how these forest lands are undermined by other policies. For instance, many 'reserved forests' are now being set aside to meet the needs of national defence, communications, industry, and other purposes of 'public importance.' The activist-educators explain that *adivasis* have no legal rights in 'reserved forests,' and that the land/forest under this classification is on the increase, given the expansionist development policy of the state. For example, 'reserved forests' are being exploited for revenue by the Orissa Forest Corporation, and by logging contracts for cheap leasing licenses on non-timber forest products that give private industry control. When the Orissa Forest Cooperation exploits forests for revenue, displaced *adivasis* in the area are directed to forest colonies that have been established to provide an ever-ready reserve labour force for the commercial projects. They cite the example of bamboo being harvested by private companies to feed paper mills in Orissa, to explain why Kondhs, like themselves, are being marginalized farther into the remote hill tops for *podu,* and having to resort to *nevad*. They explain that this is why the forests are disappearing and why *adivasis* are being denied or provided with only restricted access to forests, land, timber, and forest products.

The ideas of land classification and government ownership of land often prompt a discussion about 'outside' practices with respect to land/forest use/ownership and the 'Kondh way.' Kondh conceptions of land ownership and distribution are marked by an emphasis on community ownership (although private ownership of small plots is a recent phenomenon). Labour is shared as a community resource and the sale of land is not an acceptable practice, except in situations of extreme hardship. When a Kondh clears a portion of forest and makes the required sacrifices to the hill deity, the village holds it to be his land—but this individual makes an oath that 'if this land is sold, may the buyer/seller be smitten down with leprosy,' while beating a *tursika* branch on the ground (a bush with a leprous-lumpy exterior). 'Place' (as opposed to 'ownership') is recognized with the power of the spoken word.

The Kondhs often conclude that *nevad* and hill-top *podu* cultivation is the only way out for them, and that they should keep their receipts for fines in the hope that this might provide proof of 'ownership,' should/when the government regularize *nevad*. *Nevad* is seen to be an act of resistance against a government that Kondhs say came after *Darni* placed them on the land. So how can they take the land away from us? *Nevad* is also seen as an act of subsistence as people must necessarily encroach to stay alive. The Kondhs acknowledge the need to continue such cultivation even at the risk of state retribution because this is also a matter of material survival and subsistence.

Critical Action and Social Change

When the dialogue turns towards doing something about this situation, activist-educators and Kondh *adivasis* discuss the need to work together, to organize for strength in numbers, and to meet challenges through the traditional sense of 'community' and 'joint action.' Linking with the spiritual foundations of the Kondhs, the activist-educators invoke the powerful spiritual appeal of the need to maintain the balance in relationship between the gods and ancestors, the living community, and the natural world. They agree that, in addition to maintaining the necessary ritual observances, collective action is necessary to address 'new diseases and illnesses' of the time (destructive development) in order to regain the harmony that has been lost.

Organizations in sixty villages, grouped into six regional *fora* of ten villages, and a multi-regional forum have helped to build a network of solidarity since 1995. Collective decisions have been taken to extend *nevad*, pay fines (not bribes), maintain receipts/records, lobby the local government bodies for 'regularization' and appear at the revenue courts to gain legal recognition of the necessity to 'encroach on vacant state land/*anawadi* land,' a practice that has been used with some success. *Pattas* (land title) over their hutment areas has been secured for over 400 families, thus improving their chances of avoiding forced evictions. Collective action and inter-village cooperation has staved off repeated attempts by the state to re-appropriate land. *Nevad* under cultivation and aeforestation with fruit-bearing trees has grown eightfold over these five years. With an increasing sense of security, conservation measures such as levelling and bunding plots on the hilltops are becoming more prevalent.

These Kondhs are gradually overcoming their fear of the *sarkar* (state) as they realize that change through asserting themselves, as opposed to constantly trying to survive within the shrinking parameters of an unjust system, is painful but possible, and that they can engineer these possibilities by relying on a long tradition of collective action, participatory leadership, and consensual decision making. Those Kondhs who have had some success with land title and aeforestation jokingly speak of how revenue inspectors and forest guards who once came and made threats have now either disappeared or, if they do make an appearance, seem to speak with a 'new tongue' that is 'humble' or 'sour,' but never threatening or demeaning as in the past. As one Kondh youth said, 'We have found our voices and they seem to have lost theirs!'

Critical Reflections

Recognizing the 'Environmental': Critical Dialogue and Internal Contradictions

While popular education generates discussions about structural power and its relationship to the marginalization of groups such as the Kondhs, environmental popular education is meaningful only if the abuse of the environment by all social groups is considered. The *adivasis* activist-educators are conscious of the need to critically analyze both large-scale state-led degradation of land and forests, and the state-induced compulsions that have led *adivasis* to 'rationally' exploit the land in an all-out effort to survive. The impacts of both actions, though vastly different in magnitude and prompted by different causes, still amount to environmental destruction. While acknowledging the risk of 'blaming the victim,' critical dialogue ought to consider internal contradictions in *adivasi* environmental consciousness on the one hand and some destructive resource use patterns around *nevad* cultivation, for instance, on the other.

The activist-educators could choose to ignore the environmental consequences of *nevad* cultivation. After all, the Kondh are using already degraded forest, cultivating friable hill soils on which sustainable production could potentially be obtained. And understandably, the Kondh cannot risk the considerable money and effort required for aeforestation, bunding, and field levelling due to the constant dread of eviction. However, the end result is continuing deforestation and soil erosion. Critical action to deal with this contradiction has to address structural causes (such as pressuring the state to grant land tenure, etc.). Nevertheless, local initiatives must also be marshalled to protect the environment, using community resources and the resources of 'outsiders.' Similarly, critical dialogue to analyze the payment of bribes to 'protect encroachment' needs to take into consideration the fact that, as 'partners in crime' with the lower reaches of the state, *adivasis* are forced to ignore environmental theft and destruction when the same forest guards and revenue officials collude with private contractors to illegally clear and sell timber from 'protected forests.'

While recognizing the 'victimization' of *adivasis* by state-led development and power (that is, unsustainability is not inherent in *adivasi* practices but results from their history of political and economic domination), it should still be possible to address the destructive environmental practices of *adivasis* under duress in order to stem environmental destruction. If environmental popular educators fail to engage in such critique, *adivasis* will unwittingly become a part of the same process that is eroding the spiritual and material basis for their existence.

Recognizing the 'Popular': Adivasis as Agents of History

The eagerness of popular educators and environmentalists to incorporate *adivasis* into the discourses of environmental movements and sustainable development may arguably be "blinded by the glare of a perfect and immaculate consciousness" (Guha 1988, 84). It is also possible that their desire to construct Kondh resistance as part of a campaign for 'sustainability and social justice' could amount to an act of appropriation if it "excludes the *adivasi* as the conscious subjects of their own history and incorporates them as only an element in another history with another subject" (84). It is not possible to both hail *adivasi* as agents of history and dismiss their consciousness and the 'imperfections' of their everyday lives. The idealization of Kondh life in attempts to demonstrate that a critique of development actually exists in the lives of *adivasis* makes caricatures out of subjects. More significantly, such theoretical treatment reduces their problems and fails to recognize that their ability to mount a critique has been vastly eroded by their subordination. Freire warned of the two risks of elitism and *basism*, when he cautioned that the rejection of popular knowledge and practices was as dangerous as its exaltation or mystification (Schugurensky 1998).

The Transformative Potential of Environmental Popular Education and *Adivasis* Action in Small Localized Movements

With the help of the partnership and environmental popular education, the Kondh *adivasis* in the sixty partner-villages have succeeded, and continue to succeed, in asserting themselves and in transgressing and redefining power relations in their space. They now control or are in the process of controlling more land/forests; they have succeeded in using state laws and the justice and administrative systems to meet their interests; and they are finding a 'voice' in interactions with outsiders and state authorities. Such actions have encouraged a newly found sense of independence and a confidence in exercising their cultural option. Aeforestation and conservation measures are gradually being undertaken and, while *nevad* is expanding, a growing sense of confidence is encouraging many to reconsider traditional conservation practices in keeping with their spiritual understanding of the environment.

These changes have been realized with the help of the environmental popular education process. Nevertheless, these Kondh *adivasis* and the agents of the state are far from realizing the Habermasian notion of the "ideal speech situation" and "democratic dialogue." The ground reality of the power that divides them makes the prospect of "truth without power" utopian at worst and illusive at best. Indeed, the many examples of state repression in the face of *adivasi* assertion have not been considered here. It suffices to say that the power of the state and dominant classes and castes should not be underestimated by overstating the case for the transformatory potential of *adivasis* acting in a small localized movement. Support from 'sympathetic outsiders' and activists is crucial for realizing any transformatory potential of such movements. Furthermore, re-engagement in political struggles aimed at controlling state power at the local level, as opposed to steering clear of the state/party-political process, might be necessary. *Adivasi* movements can engage in pragmatic action that is both cooperative and confrontational, as and when it is appropriate, based on and guided by their rationale for existence.

In our attempts to focus on the transformative potential of *adivasi* movements in dealing with the concerns of nature and social justice, environmental popular educators need to recognize that the burden of history and the environment cannot be placed on the shoulders of *adivasis* alone. Strengthening the struggle to preserve the *aranya sanskriti* of the *adivasis* addresses the social justice and environmental concerns of helping enable the Kondhs to sustain themselves and their cultural reality, and places the forests in the hands of those that leave the smallest 'ecological footprint' (Sachs 1997).

However, transformation of the social structures that sustain the process of modernization and development will require more than a simple reliance on *adivasi* movements 'to do the job for us.' In order to alter the course of history and transform the power of the state and the market in the service of humanity and the environment, it is necessary to combine the global power of social movements that profess even a partial common interest in establishing natural and/or social limits to global development, modernization, and progress.

References

Baviskar, A. 1995. *In the Belly of the River: Tribal Conflicts over Development in the Narmada Valley.* Delhi: Oxford University Press.

Chambers, R., Saxena, N., and Shah, T. 1989. *To the Hands of the Poor: Water and Trees.* London: Intermediate Technology Publications.

Clover, D. 1995. "Theoretical Foundations and Practice of Critical Environmental Adult Education in Canada." *Convergence 28,* 4, 44–51.

Foucault, M. 1980. "Power-knowledge." In C. Gordon (Ed.), *Selected Interviews and Other Writings: 1972–1977.* New York: Pantheon Books.

Freire, P. 1970. *Pedagogy of the Oppressed.* New York: Continuum.

Freire, P. 1994. *Pedagogy of Hope: Revisiting Pedagogy of the Oppressed.* New York: Continuum.

Gadgil, M. and Guha, R. 1992. *This Fissured Land: An Ecological History of India.* Delhi: Oxford University Press.

Giddens, A. 1984. *The Constitution of Society.* Cambridge: Polity Press.

Guha, R. and Gadgil, M. 1989. "State Forestry and Social Conflict in British India." *Past and Present: A Journal of Historical Studies, 123,* 141-177.

Guha, R. 1988. "The Prose of Counterinsurgency." In R. Guha and G.C. Spivak (Eds.), *Selected Subaltern Studies.* New York: Oxford University Press.

Habermas, J. 1984. *The Theory of Communicative Action: Vol. 1.* Boston: Beacon Press.

Habermas, J. 1987. *The Theory of Communicative Action: Vol. 2.* Cambridge: Polity Press.

Hall, B. and Sullivan, E. 1994. "Transformative Learning: Contexts and Practices." In *Awakening Sleepy Knowledge.* Toronto: Transformative Learning Centre, OISE.

Inglis, T. 1997. "Empowerment and Emancipation." *Adult Education Quarterly 4,* 1, 3–17.

Nayak, R., Boal, B., and Soreng, N. 1990. *The Kondhs: A Handbook for Development.* Delhi: Indian Social Institute.

Orr, D. 1992. *Ecological Literacy: Education and the Transition to a Postmodern World.* Albany, NY: State University of New York Press.

Oxfam, 1996. *Oxfam Orissa Strategic Plan.* Bhubaneshwar: Oxfam, India.

Sachs, W. 1997. "The Need for the Home Perspective." In M. Rahnema and V. Bawtree (Eds.), *The Post-Development Reader.* New Jersey: Zed.

Schugurensky, D. 1998. "The Legacy of Paulo Freire: A Critical Review of His Contributions." *Convergence 31,* 1 and 2, 17–29.

Shiva, V. 1989. *Staying Alive: Women, Ecology and Development.* London: Zed.

Shiva, V. 1993. *Monocultures of the Mind: Perspectives on Biodiversity and Biotechnology.* London: Zed.

Shiva, V. and Bandyopadhyay, J. 1990. "Asia's Forests, Asia's Cultures." In S. Head and R. Heinzman (Eds.), *Lessons of the Rainforest.* San Francisco: Sierra Club Books.

UNDP. 1992. *Human Development Report.* New York: Oxford University Press.

Welton, M. 1993. "Social Revolutionary Learning: The New Social Movements as Learning Sites." *Adult Education Quarterly 43,* 3, 152–164.

Part II

Women's Knowledge, Education, and Activism

Chapter 4

Public Space and the Aesthetic-Politic of Feminist Environmental Adult Learning

Darlene E. Clover
University of Victoria, Canada

Introduction

> The ritual and building of experiences in place is what creates poetry in the landscape. (Mongard, 1999, 28)

> Take the world as is, not as it should be—then move it to where it should be. (Homan, 1999, 207)

Feminist community arts are not and have never been insignificant tools of struggle. Women's art, and the process of art-making, can be powerful catalysts to stimulate imaginative thought, critical dialogue, community mobilization, personal transformation, and socio-environmental change (for examples, see Clover 2000; Roy 2000; Stafford 1999; Stien, 1994; Withers 1994). Feminist arts projects may be individual or collective. They may celebrate and beautify, condemn and reveal injustice, controversy, and inequity, or actually do both at once. But as they "suggest a new way of looking at the world and change our way of looking at ideas and norms," they are never neutral (Dodd 2000, 2).

Cameron (1989, 64) argues that there has always been a strong link between feminism and the environment. From its beginning, it has been

> actively involved in the peace movement, in the anti-nuclear movement, and in the environmental protection movement. Feminism is what helped teach us that the link between the political and industrial included the military and was a danger to all life on the planet.

This chapter examines two feminist aesthetic environmental adult education projects. While differing in their locations and approaches—one project is situated in the urban environment and the other in a rural setting—both projects tap into the imagination and challenge the notion of 'public space,' in which "everyone is a stakeholder and has a sense of proprietary—and where conflicts of interest and usage are inevitable" (Felton 1999, 6). Although the idea of 'place' is often difficult to define, it is something which is not neutral. As Helen Broadhurst (1999, 34) argues, "every place, no matter how 'godforsaken' to an outsider, engenders passion in the people actually living there." Things and events that matter, that have the potential to reshape people lives and form new trajectories, most often occur within place, "the immediate environment and the community" (London 1994, 4). Public space is valued in particular for the important role it plays in creating a communal experience (Mongard 1999).

The community arts project titled "In the Hood" took place in a low-income, culturally diverse neighbourhood in the city of Toronto. Community arts are a collaboration between professional artists and community members to advance artistic and community goals, using participatory and creative processes. From creation to completion, the work is guided by a collective vision. In this collective process, feminist artist-educators used traditional craft making to transform the spirit of a community and the urban landscape.

The second arts project titled "The Developers' Feast" took place on the southernmost tip of Vancouver Island, British Columbia, in the neighbourhood of Metchosin. Challenged by a collective of women artists who meet on a monthly basis, Gretchen Markel created a visual icon to celebrate the beauty of the rural landscape, to visually preserve that which needs to be remembered. But the artwork also has a very critical dimension. It poignantly juxtaposes the beauty of the landscape with the destructive practices of unharnessed development and, by doing so, creates controversy wherever it goes and stimulates debate and dialogue.

In the Hood

> While the city...is often seen as 'too big, too undefined' the locality can be described, quantified and experienced. In the global city the locality, like a bright kid in a dysfunctional family, is capable of redemption. Within this possibility the process of community revitalisation is position—the local both recovered and produced. (Guppy 1999, 9)

In the urban environment, a "sense of place is as much about the bricks and mortar of built form as it is about our imaginative connection with it" (Felton 1999, 8). While some urban neighbourhoods prosper and maintain their beauty and liveability, others deteriorate to empty shops and glass and litter-strewn pavement. As the deterioration of the physical community grows so do feelings of helplessness—that people have little or no control or say over the future—and of fear, isolation, and alienation. This is problematic for, as London (1994, 4) suggests, more than any other it is the local, the community, that "is the arena for the creative expression of personal encounters with one's environment, one's web of life...that inextricably embraces, defines, and empowers."

"In the Hood" was a community arts project located in a culturally diverse neighbourhood in central Toronto, Oakwood and Vaughan. A fitting framework for the project is a quotation by Aroko (in Farkas 2000, 15), who argues that "while art cannot completely solve many social dilemmas, it can solve one problem: The loneliness of spirit." In spite of living amongst thousands of people, cities are often impersonal spaces in which feelings of alienation and isolation are found. Enhanced movement toward the 'individual' and the resplendent uselessness of a private life centred on nothing but itself and away from the collective, the meaningful, and in particular the imaginative, leaves people feeling helpless, hopeless, detached, and uninspired (Greene 1995). This is problematic as it is the imagination that most effectively enables us to cross the empty spaces between ourselves...and others. This is because, of all our cognitive capacities, imagination is the one that permits us to give credence to alternative realities. It allows us to break with the taken for granted, to set aside familiar distinctions and definitions (Greene 1995).

The multi-ethnic neighbourhood of Oakwood and Vaughan is characterized by high unemployment and crime. Shop owners often find it more profitable, as an income tax deduction, to leave their shops empty rather than fill them with low-budget or secondhand shops. A street with many empty shops and few people looks derelict, sad, and empty. As articulated by artist-educator Elizabeth Cinello, in an interview in 2001:

> The boarded-up shop front windows in the neighbourhood are an interesting metaphor for the community itself. What normally is wide open and designed to facilitate seeing, is instead covered, hidden and neglected. The richness, diversity, and colour of the hood's culture and environment are unseen, ignored and invisible, except as a stereotype in the media.

In 2000, feminist artist-educators went door-to-door to talk with people about their neighbourhood and discovered three very important factors. The first was that, although many women created crafts—they knit, sewed, or crocheted—these activities were most often kept to family or friends, the 'private' domain. Walker and Walker (1987, 27) argue that "craftwork performed by women within a domestic setting has been done for love and for creative need, and also for domestic economy: it is an aesthetic survival." For centuries, "women have communicated through craft activities, developed relationships and have been transformed" (Larvin and Pooley 1987, 11). Although the word 'art' began as a representation of craft, women's crafts have more often been relegated to the realm of mere 'hobby' and 'decoration.' Secondly, as noted by the feminist artist-educator Elizabeth Cinello, although numerous arts projects were actually taking place around the community, "greater interaction needed to take place between different groups in the community to support one another." Thirdly, while people had very mixed feelings about their community, they appeared to be hungry for something to do, to get involved, and all that was really required was for someone to take the lead.

Based on these findings, the artist-educators designed a community arts project that could deal with two major issues. One issue was isolation and the feeling of disconnection. As people become more isolated from one another, there are fewer opportunities to engage in the informal learning processes that contribute to the development of collective knowledge and social change. The second issue was to make visible women's creative activities, as artistic practices. The third was to use people's skills and their crafts wherever possible to transform an ugly, drab, and soulless streetscape. In an interview, artist-educator Jo Anne Atherly described the purpose of the project this way:

> [It] is about bringing art to the community; about transforming our built environment, about investment in the arts and each other; about being able to recognise each other on the street; about recognising our different talents; about us.

In short, the primary goal of the project became getting people out of their homes and valuing their creative energy. According to one of the project participants, "[creativity] is a necessary tool we all need to explore. If we look at having some hope for future healthy communities we all need to tap into those creative energies. Imagination can transform this place. It has already if you just look at the laundromats."

Through a series of community workshops in the library, a variety of artistic expressions were created. One was a giant collectively knitted slipper. Each of the women involved knitted a small piece which was then woven into a large, multi-coloured slipper. Collaboration flowed from the rapport of the women's interactions, and the designs and ideas triggered sparks that influenced the flow and nature of the collective piece. The second activity were small individual fantasy slippers into which personal stories and dreams were woven. Once this phase had been completed, the artist-educators lobbied shopkeepers throughout the neighbourhood, requesting that, if they had no intention of using the shop front windows, they would be best filled with community art. And, in many cases, it worked. A third activity was the transformation of one laundromat's unsightly and cavernous interior into a work of art through poetry and paint. One of the women involved in the project said that she had actually been afraid of the laundromat until the transformation took place. Transforming space into something more inviting and safe is extremely important because as Felton (1999, 7) argues:

> one of the fundamental conditions of urban life is the close proximity in which we all live together among strangers.... For women, the reality of living among strangers affords along with the opportunities, a different set of possibilities where sexual difference is highlighted and might also work to limit potential rather than enhance it...the negotiation of space in a city of strangers requires an act of trust and discretion on women's part and an assumption that encounters with male strangers will be benign and not malevolent.

A fourth activity was the creation of re-usable cloth "designer laundry bags" that were adorned with words, symbols, found objects, and imagination. People passing by the laundromat saw these designer bags on display and began stopping in to see if they could purchase them. A small cottage industry was born. The final activity was to go to the schools and encourage children to come to the neighbourhood centre *ARTSTART*, after school, to do some arts projects. When the artist went into the school, she thought there might be a few children interested in coming to the centre; she announced it over the PA system. All the children, as well as the teachers, showed up in the office wanting to take part. A most interesting comment came out of this activity, which is articulated by the artist-educator, Starr Jacobs:

> While the kids were creating they were also talking about things. Sometimes their remarks were not good. I said to one girl "Why do you keep calling him [her

brother] a faggot? Do you know what a faggot is?" This little girl was particularly bad but we were able to talk to her. This creative space made her more able to listen to what we were saying to her. Kids learn these things at home and in the schools and community. You have to deal with it and sometimes it's like being a counsellor. Sometimes you don't know what to say or do. I guess it's better that these things come out because you can deal with them. This project really gave me a window into what kids today are picking up and how art can be used to help them work through it.

The more things exhibited around the neighbourhood, the more the adults' and the children's relationships to 'place' and their ideas of their crafts were transformed. The more children hooked up with adults, the more they saw themselves as part of the 'community' and as people who had something to contribute.

Rehabilitating a small part of the community does not create the conditions for change that would eventually give people the ability to manage the infrastructure and avoid future decay. However, the project did achieve a number of extremely important things. First, it helped to overcome feelings of isolation and disconnection. As one woman noted "I live alone, you see. And I get lonely. You can't just watch TV all the time." Second, it engaged people's creativity and imaginations and helped them to turn a few elements of their built environment into something more beautiful. Third, it gave the participants an opportunity to work with local artist-educators and see themselves as artists. As one woman noted,

Now that I have learned from the artists...I would plan the theme for my slipper better. I would do two slippers—one to express the impromptu, untamed side of me, and the other a more 'cultivated slipper' to express the calmer, aesthetic and intellectual side.

Fourth, the women gained confidence around their own skills in knitting, sewing, and crocheting to begin to teach them to others. Fifth, the small changes enabled the women to see themselves as agents of change, as people who, by working together, could actually make a small transformation in their urban environment. Finally, this aesthetic learning process highlighted the cultural practices of the neighbourhood, improved the streetscape, raised one aspect of women's everyday lives to an 'art.'

The Developers' Feast

To breathe in means to breathe out. What artists feel deeply must 'come out' in their artworks and communities. Good art is art of consequence, and those who make art of conscience and consequence are engaged in activism. (Ressler 1996, website)

Of primary concern in the rural areas is an escalation in the removal of public land, space, and resources from the hands of the public and into the hands of private developers and speculators. The argument goes that the private sector can deal more 'efficiently' with these under-utilized spaces and resources (such as parks), and that there is a 'desperate' need for large-scale development projects and their 'positive' contributions to the economy. However, the "power and politics in the landscape is often an abuse. Those with influence build what they desire" (Mongard 1999, 28). Much of what takes place under the guise of 'development' is mired in disrespect, greed, destruction, and political manipulation. Moreover, the 'economic' arguments appear never to take into account the spiritual need that people have for green spaces, the need to preserve habitat, and the need to conserve resources for the future. As Sister Mary, a nun from East Sooke engaged in a struggle against a so-called eco-tourism development project, questioned, "What right do they have to make money at the expense of everyone else?"[1]

Felton (1999, 7) argues that, for a number of decades, feminist planners and geographers "have drawn attention to the contested nature of public space. In particular, the fact that space is a site and source of social power and an arena where social relationships are meaningful." Moreover, during the 1970s and 1980s, women artists "began to conceive of a culture that would revolutionize our relationship to [the rest of] nature" (Feman Orenstein 1990, 281). Working within an environmental framework, feminists began to use the arts to encourage a union with the Earth, provoke different ways of seeing and understanding landscape and space, and communicate the gendered politics of environmental problems (Felton 1999; Stein 1994). The artwork of women spiritually nourished by the Earth is intimately fused with connections to the rest of nature.

"The Developers' Feast" was created in Metchosin on Vancouver Island. It is an area of mountains, lakes and streams, old-growth forest, rich agricultural lands, an abundance of marine and land-animal life, and a mediterranean climate. However, with the virtual collapse of natural resource industries, such as fishing and logging, tourism is the fastest growing industry and with it comes development and more development.

Although the municipality of Metchosin struggles and votes against mega-development in the area, the surrounding towns of East Sooke, Colwood, and Landford have homogenized and paved their landscapes with American-style mega-malls and fast food chains.

Arguing that "we have been fighting [development projects] in the traditional ways and are getting nowhere," a group of women artists from Metchosin encouraged one of their own, the one seen to be the most environmentally focussed, to use her art to document and challenge ongoing or planned destructive development projects in their area, and visually capture their impact—the before and after—on ecologically sensitive areas. The result was the creation of "The Developers' Feast," a life-size picnic table of stunning beauty which is both an ecological memory and an ecological present. Feminist artist-educator Gretchen Markel chose the idea of a feast because she felt it best captured the ways in which "development was eating up the land" and the wild places "were being dished up to development."

Because the issues are so large, contentious, and frightening, the feminist artist wanted to create something humourous yet poignant that could be taken to public meetings, to encourage discussion about the blatant handing-over of public land by government to private development schemes. People need to have the opportunity to debate whether or not the individual should precede the community and use their imaginations to regain the public spaces being lost.

Running down the centre of the picnic table is a silk tablecloth which is frayed at one end to show what has been lost, but also to show, as the feminist artist notes, "how we are all feeling these days—a bit frayed and frazzled around the edges from our years of battle." The end where the cloth is not frayed represents glimmers of hope. The silk cloth is covered in red hearts or "bleeding hearts," as the women are often referred to, by politicians and developers with whom they do not see eye-to-eye.

The knives are miniature chainsaws as 'clear cutting' is the practice of choice on Vancouver Island. Some of the plates contain drawings of pristine areas while others include drawings of the problem of development such as clear cutting. To create the drawings for each plate the artist rode her bicycle around the community and made sketches. In the centre of the table are menus that contain humourous yet truthful accounts of the 'befores' and 'afters' of development.

One area painted on a plate is Matheson Lake. Although it is located in a provincial park used by thousands of people, a developer managed to obtain permission to log the site and the trees are coming down. So the menu reads:

Beverages: Accompany your fine meal with a tall glass of sparkling, cool, clear Matheson Lake Water. Caution: Our apologies, Madames et Messieurs, but recent cutting in the Matheson Lake watershed, courtesy of [the developer], has compromised the purity of this beverage.

Although the scene on the plate is one of degradation, it is still one of beauty. The feminist artist noted that the reason for this is that, even in its pain, "the land is beautiful and still maintains a certain dignity."

Another plate depicts the 'Galloping Goose Trail,' named after a historic steam engine, and built upon an old railway line. A major gas company lobbied local authorities to run their gas pipeline down the Galloping Goose and, as another plate depicts, the trail is being dug up. So the menu reads:

Appetizer: Galloping Goose—Long strips of delicate woodland, sprinkled with wildflower blossoms, to be seared over…gas and seasoned with other secret utilities.

Another plate carries a painting of East Sooke Park, a wilderness area of mostly rock richly covered in mosses and other forms of life. Beside East Sooke Park is private land that has been purchased by a developer. The developer's plan is to create an eco-tourism destination resort. The resort would include a hotel with 85 rooms, 15 cabins, a nine-hole golf course, and a marina with 115 births, as well as an approximately 200-home subdivision with a new waterline. Many studies have shown that this project will endanger cormorant nesting sites and fish habitat and contribute greatly to the pollution of the Sooke Basin. The menu reads:

East Sooke Park: Magnificent *cote-de-mer*, a fine blend of Douglas Fir and Hemlock zones rich in seafood and greens. Served with a side dish of stripped Silver Spray.

This development project has torn the small community of East Sooke apart. Some people in the area see it as a source of new jobs and something that will bring a more stable water supply to the area. Others feel it will irrevocably alter their rural way of life and destroy the environment. Lawsuits abound, legal fees are in the thousands of dollars, and heated public meetings run into the small hours of the morning. In spite of the lawsuits and numerous threats that have been made against private citizens, the residents press on, taking part in meeting, after meeting, after meeting. They have managed to stave off the development project for over four years.

Erica Dodd (2000) argues that art, as it suggests "new way of looking at the world [changes] our way of looking at ideas and norms." In this sense, as is true with The Feast, it can be a warning and a comfort. As a 'warning,' The Feast is brought to community hearings and meetings on development projects and often, due to its 'controversial' nature and the challenge it represents, has had to be left outside the meeting. As a 'comfort,' the artwork won the People's Choice Award at a local arts festival entitled "Celebrating Green Spaces," in 2001. As the only 'political' piece of art entered, the award demonstrated how hungry people in the area were to see the issue of development brought to the fore. The Feast was able to reflect issues of vital importance in the community, and to raise consciousness about the environment in a way that no other language could achieve (Dodd 2000).

Conclusion

> When interest [in emancipation and the aesthetic] is aroused and fulfilled, people seek and find greater freedom from their oppressive domination. They do this through self-reflective discourse and artistic expression. (Harris, 1999, 110)

Art, broadly defined, is "that which is an integrated manifestation of feeling, idea and form" (Sherk, in Withers 1994, 158). It creates and engages us in a language of sight, symbol, sound, texture, and metaphor. The arts are often "richer in meaning when they are an expression of experience that matters deeply" (London 1994, 4). That way they allow for a

> three-dimensionality of experience in that they challenge taken-for-granted ways of seeing and being in the world and, in so doing, they provide an 'aesthetic shock' that propel[s] people to question everyday assumptions. (Harris 1999, 110)

As the two projects demonstrate, feminist artist-educators use the arts for a variety of environmental learning purposes. They use them to make visible the invisible of women's experiences—their knowledge, skills, and creativity. They use them to celebrate the urban environment, that which is beautiful, that which remains, while they also challenge normative views and transform space. Feminists working with and through the arts find creative ways to express human relationships to the rest of nature and enhance learning through questioning and the connection of the imagination. For it is the imagination that "permits us to give credence to alternative realities...[and] allows us to break with the taken for granted, to set aside familiar distinctions and definitions" (Greene 1995, 3). The imagination provides the catalyst for people to resist state and corporate violence, seek social and environmental redress, "and design new forms of civic engagement and collaboration" (Greene 1995, 6).

The two aesthetic feminist environmental adult learning projects examined through this chapter worked to facilitate a deeper understanding of the complex and multi-layered challenges within urban and rural environments. The artist-educators provided ways to promote consciousness and imagination, and to communicate important political, social, and environmental concerns. In essence, the projects are very much about [the]

> realization that the individual does not precede the community.... More and more of us...are aware of how necessary it is to keep such visions of possibility before our eyes in the face of rampant carelessness and alienation and fragmentation. It is out of this kind of thinking...that the ground of a critical community can be opened.... It is out of such thinking that public spaces may be regained.... The principals and the contexts have to be chosen by living human beings against their own lifeworlds in the light of their lives and others, by persons able to call, to say, to sing, and— using their imaginations, tapping their courage—to transform. (Greene 1995, 198)

Note

From a Canadian Broadcasting Corporation (CBC) news programme, 2001.

References

Broadhurst, Helen. 1999, Edition 3. "Butterfly Power—Working with Chaos." *Queensland Community Arts Network News*, 34–37.

Cameron, Anne. 1989. "First Mother and the Rainbow Children." In Judith Plant (Ed.), *Healing the Wounds: The Promise of Ecofeminism*. Toronto: Between the Lines Publishing, 54–66.

Clover, Darlene. 2001, March. "Aesthetic Activism: Women and Community Arts." *Artwork Magazine*, Community Arts Network (South Australia), 1–7.

Dodd, Erica. 2000. "The Universal Language of Art in the Multi-Cultural Society." In Ludgard De Decker (Ed.), *Art as an Early-Warning System*. Victoria, BC: University of Victoria, Centre for the Studies of Religion and Society.

Farkas, Suzanne. 2000, Summer/Fall. "Women Artists Creating Space for Healthy Communities." *WE International*, 15–17.

Felton, Emma. 1999, Edition 3. "Women's Place in the City of Postmodernity." *Queensland Community Arts Network News*, 6–8.

Feman Orenstein, Gloria. 1990. "Artists as Healers: Envisioning Life-Giving Culture." In Irene Diamond and Gloria Feman Orenstien (Eds.), *Reweaving the World: The Emergence of Ecofeminism.* San Francisco: Sierra Club Books.

Greene, Maxine. 1995. *Releasing the Imagination: Essays on Education, the Arts and Social Change.* San Francisco: Jossey-Bass Publishers.

Guppy, Marla. 1999, Edition 3. "The Edge of Centre: Local Identity and the New Neighbourhood." *Queensland Community Arts Network News,* 9–15.

Harris, Carol. 1999. "Passion and Politics: 99 Years of Adult Education." In M. Hrimech (Ed.), *Proceedings of the 18th Annual Conference of CASAE.* Montreal: University of Montreal, 109-113.

Homan, Mark. 1999. *Rules of the Game: Lessons from the Field of Community Change.* Toronto: Brooks/Cole Publishing Co.

Larvin, Jenny and Pooley, Sue. 1987. "Our Family Heritage: A Conversation Between Two Sisters." In Gillian Elinor, Sue Richardson, Sue Scott, Angharad Thomas, and Kate Walker (Eds.), *Women and Crafts.* London: Virago Press.

London, Peter. 1994. *Step Out-Side: Community-Based Art Education.* Portsmouth, NH: Reed Publishing.

McFee, June and Degge, Rogena. 1997. *Art, Culture and Environment, A Catalyst for Teaching.* Belmont, CA: Wadsworth Publishing Company, Inc.

Mongard, John. 1999, Edition 3. "Power, Poetry and Contemporaneity in the Public Landscape." *Queensland Community Arts Network News,* 28–39.

Overton, Patrick. "The Role of Community Arts Development in Nurturing the Invisible Culture of Rural Genius." In *From Artspeak to Artaction: Proceedings of a Community Arts Development Conference.* Saskatoon, Canada: The Extension Division, University of Saskatchewan, 87–97.

Ressler, Susan. 1996. "It's All About the Apple, Or Is It?" *Women Artists of the American West.* http://www.sla-purdue.edu/waaw/Ressler/Ressleressay3.html

Roy, Carole. 2000. "Raging Grannies and Environmental Issues: Humour and Creativity in Educative Protests." *Convergence XXXIII,* 4, 6–17.

Stafford, Eve. 1999. Edition 3. Rural Women for Cultural and Community Leadership. *Queensland Community Arts Network News,* 30–33.

Stein, Judith E. 1994. "Collaboration." In Norma Broude and Mary Garrard (Eds.), *The Power of Feminist Art.* New York: Harry N. Abrams, Inc.

Walker, Agnes and Walker, Kate. 1987. "Starting with Rag Rugs: The Aesthetics of Survival." In Gillian Elinor, Sue Richardson, Sue Scott, Angharad Thomas, and Kate Walker (Eds.), *Women and Crafts.* London: Virago Press.

Withers, Josephine. 1994. "Feminist Performance Art: Performing, Discovering, Transforming Ourselves." In Norma Broude and Mary Garrard (Eds.), *The Power of Feminist Art.* New York: Harry N. Abrams, Inc.

Chapter 5

Women, Literacy, and Environmental Adult Education in Sudan

Salwa B. Tabiedi
USA

Introduction

Sustainable development must be women-centred. I am not saying this because I am
a woman or a feminist. I am saying this because historically, and even today,
women most often take care of the basic needs of society like food, fodder, fuel,
shelter, and nurturing. (Bhasin 1992, 32)

The United Nations Conference on the Human Environment, held in
Stockholm, Sweden, in 1972 was a turning point in the upsurge in efforts to
address global concern for the environment. While scientists argued with
policy and decision makers about the possible catastrophic effect of present-
day environmental degradation and natural resource(s) depletion and the
need for change, educators suggested that environmental education could be
a major tool in bringing about change if it was re-conceptualized and
broadened in its vision and scope. The United Nations Environmental
Programme (UNEP), and the United Nations Educational, Scientific, and
Cultural Organization (UNESCO) have organized conferences, workshops,
and documentation projects in response to this challenge (Ahmed 1991).

Much progress has been made in incorporating environmental issues and
awareness into school educational systems but support for environmental
adult education remains disturbingly low. As environmental problems
continue to escalate in the Sudan, it is becoming obvious that environmental
education for children alone is insufficient. It is of vital importance that
quality environmental education for adults be developed and made available
to all, whether literate or illiterate, male or female, urban or rural. It must
also be made available to those who do not and probably will not ever again
attend educational institutions. As a result environmental adult education has

developed in Sudan to enhance human quality of life through the conservation of natural resources. Environmental adult education is particularly important as adults are the most active agents of change in communities and the primary educators in their families.

Through the lens of some of the most pressing socio-environmental problems in Sudan, this chapter explores some of the basic elements of environmental adult education, especially with, by, about and for women, and a number of activities taking place. The chapter also looks at women's literacy, calling for environmental adult educators in Sudan to explore contemporary practices in environmental adult literacy which might prove beneficial to their work.

Major Environmental Problems in Sudan

> Where drought is the epic then there must be some
> who persist, not by species-betrayal
> but by changing themselves
> minutely, by constant study
> of the price of continuity
> a steady bargaining with the way things are.
> (Rich, The Desert as Garden of Paradise 1989)

Sudan has a number of environmental problems that demand immediate attention, but the single biggest problem is drought. Primarily caused by desertification, years of drought have led to famine, and enormous crop, vegetation, and livestock losses, all of which have had immense social and economic costs.

A series of droughts occurred in 1913, 1968–73, and 1983–85. These have affected all of Sudan but areas such as Darfur, Kordofan, and parts of the Eastern Regions have experienced the most severe impacts. In 1983, the area affected by drought was 650,000 km^2 (26.5 percent of Sudan) especially in the northern and eastern regions of Sudan (Tabiedi, 1987). In 1990, Kordofan and Darfur were hit by another severe drought that led to drastic shortages of water and food (Ministry of National Planning 1991). Droughts can result from a modification or disruption of global systems or from the local effects of increased human activities that cause imbalances between population and resources. Desertification in Sudan is occurring at a rate of 6 kilometres per year (El Jack 1992, 2) and is largely caused by human misuse or over-exploitation of natural resources. It is particularly severe in areas of traditional agriculture, primarily undertaken by women.

The Primary Causes of Drought and Desertification

> Although much of the Arab Region's environment is desert, it consists of many renewable and non-renewable natural resources which affect its culture and future. (Al Agib 1992, 75)

Drought from desertification in Sudan is a result of various economic, social, and administrative factors that have exacerbated natural climatic variability. Economic and social life in the areas affected by drought depend heavily on forest resources for firewood, home and furniture building, hunting equipment, and traditional agricultural tools and weapons such as spears, axes, and knife handles. Trees have also been removed to clear the land for settlements and farming without due consideration of the impact of population pressure on the carrying capacity of the land. Furthermore, the overgrazing of vegetated pastures gradually has exposed the land to desertification.

Administrative factors also contribute to some of the problems. There is little effective supervision of forests and pastures, while the insufficient cleaning of fire lines has resulted in a loss of forests and pastures. The lack of coordination between agriculture and forestry authorities has also led to the removal of large tracts of forested land without consultation, while the absence of an effective system for curbing the smuggling of firewood has resulted in massive deforestation in western Sudan. The area of reserved forests stands at three million *feddan.* This is equivalent to about 0.5 percent of the country's area when it should be no less than 20 percent, according to land-climate-vegetation relationship.

Climatic factors are important in Sudan. Some areas receive very low rainfall, including the Savannahs (west and north of Um Ruwaba, north of Bara, north of El Nihod, and south of Sodri) and the semi-desert and desert zones north of Sodri. Here, rainfall is scarce and the soil tends to be sandy, making the land vulnerable to erosion and desertification. Civil war in the south has caused 1.5 million refugees to make their way from neighbouring countries into Sudan to escape violence, political instability, and famine. This has compounded the pressure on the few available land resources. As a result of these many factors, most of the land in Sudan used for traditional agriculture has been affected by desertification. This has caused crops to fail, productivity to decline, and food reserves to shrink, resulting in hunger and malnourishment, particularly of women and children.

Al Agib (1992, 75) suggests that education is a crucial part of any solution to these problems:

> Although many efforts were undertaken through legislation and rules to preserve and protect the environment in the Arab Region, it still remains that the best solution lies in education for individuals to the extent that it enables them to build positive values and trends towards the environment for the betterment of their lives and societies.

Agriculture in the Sudan is typically a female-dominated activity. Through their various activities, women affect, and are affected by, the natural environment that is vital for agricultural development and daily survival. Thus, though desertification affects the entire population, its effects are felt most seriously by women. This is because women come in contact with natural resources through their daily household activities such as planting, animal husbandry, handicraft production, food preparation and preservation, and the collection of water, fuel, or traditional plants for medicine.[1] This vital link between women and the environment in the context of development cannot be ignored.

Women and the Environment

> Women had knowledge and skills in agriculture, animal husbandry, crafts and medicine. But when these activities were commercialised and industrialised, women often lost out. Their knowledge was declared traditional and therefore, unscientific and redundant. (Bhasin 1992, 28)

Since the 1970s, the issue of women's participation in the economies of developing countries has gained importance. Enhancing the role of women in the development of their nations is an increasingly important national socio-economic goal. The centrality of women's roles in solving the food crisis was recognized by the Organization of African Unity's Lagos Plan of Action for Economic Development of Africa in 1980. Similarly, a review of "Forward Looking Strategies of Implementation for the Advancement of Women for the Period 1986–2000" at the Nairobi Women's Conference in 1985 stated that, as food producers, women play a central role in food production and development. As such, programmes concerning food and agriculture should be designed to reflect women's involvement and to include their activities.

Indeed, any study of the relationship between household activities and the environment reveals that both urban and rural women make the primary contact with basic resources that provide the basis of family life: water (for cooking, drinking, and washing), trees (for furniture, home building, tools, and fuel) and plants (for medicine and food). Thus, the specific relationship that women share with the natural environment through their productive and reproductive activities cannot be denied, nor can we deny their significant role in the management, planning, and improvement of degraded environments and resources.

Apart from household tasks that revolve around the collection of fuel and water, child care, preparation of food, and the maintenance of proper sanitation and waste disposal, many women in developing countries are directly responsible for the production, storage, preparation, and processing of food through traditional agriculture. In western Sudan, for example, women perform 60 percent of food storage, 100 percent of food processing, 60 percent of food marketing, and 100 percent of cooking and serving meals (Abu Affan 1989, 2).

Women are also significantly involved in animal production in the traditional sector, where small animals such as sheep, goats, and cows are kept within the household. Women share in the herding, feeding, milking, and preparation of animal by-products for consumption, and use the surplus for income generation. Thus, in Kordofan, women have plots within or near the households where farming is wholly women's activity, while in southern Sudan all activities related to the production of certain vegetables are considered women's work (El Nagar 1991, 17). Women also are most often responsible for fetching firewood and water, and making handicrafts from local materials. These many responsibilities are increasing as the number of female-headed households is growing, as men leave in search of employment in the urban areas or distant industries.

Environmental degradation means women's work has become more burdensome. For example, rain-fed agriculture, the primary agricultural activity for women, has been severely affected by desertification in the last three decades. To overcome this problem, women have been compelled to expand the area they cultivate to increase crop yield, look for new grasslands for their herds, and walk longer distances to fetch water and firewood. These activities increase the time women have to spend maintaining family subsistence. The magnitude of these problems has forced women, in many cases, to abandon their homelands and work as migrant wage-labourers elsewhere. Thus, women's participation in the labour force of western Sudan is estimated at 80 percent (Khidir and Mahmoud 1991, 7).

Environmental Adult Education with, by, about, and for Women

Environmental adult education in Sudan can strengthen the contribution of women to conservation through the many activities they perform with and within the natural environment. Providing women with access to natural resources is not enough. Allowing them to work together and develop new ideas, and ways of using and maintain these resources in an ecologically sound and sustainable way is the most important process. For example, while women need to ensure that sufficient water is collected, they also need to be able to discuss and analyze the quality of that water and the reasons why pollution is occurring (Jong-Boon 1990).

Collection of firewood by rural women for their families' use is often not considered a major cause for deforestation. However, according to Forestry Department reports in Sudan, wood resources constitute about 80 percent of fuel consumption in Sudan and 70 percent of this is used as household fuel. This means that 51 m³ of wood from a cleared five million *feddan* are used annually as fuel (Abdel Latif 1992, 4). The shortage of fuel wood can adversely affect food preparation and nutrition since food cannot be properly cooked. However, environmental adult education, which engages women in a focus of selecting the most nutritious foods using local fruits, plants, and vegetables, and developing additional sources for food, can offset this impact. Thus, environmental adult educators are working to provide women with the opportunity to share their knowledge of the relationship between the causes and effects of environmental degradation, and to use this knowledge as well as their diverse capacities as managers and educators to solve local problems. Bringing women together to exchange experiences and ideas, especially of their indigenous ecological knowledge, is an effective move towards socio-environmental change. This approach to environmental adult education is empowering for women and can enhance their position as leaders in the improvement of the environment.

Women's daily contact with their children, spouses, and other members of their families creates opportunities where women can present information, inculcate ecological ethics, attitudes, and direction, and foster positive actions towards the environment. As primary caregivers for children, women can teach their children to understand the inter-connection of all aspects of the environment through direct experience and participation in the daily activities of life. This effort will help the future generation to better understand and appreciate their role in the conservation of the environment.

Finally, women can gain much from the introduction of appropriate technologies and labour-saving methods that deal with the shortage of fuel, water, and other resources. For example, women environmental adult educators in the Joint Environment and Energy Programme (JEEP) in neighbouring Uganda work with women in their own communities to further develop their skills in fuel wood conservation technologies such as fuel-saving stoves, agroforestry in the form of tree planting, and soil and water conservation through organic farming. In Sudan, the Ministry of Energy works with some non-governmental organizations (NGOs) to develop environmentally friendly fuel-saving stoves.

Women and Literacy

> Not being literate amounts to surrendering power to literates to claim knowledge which the disadvantaged cannot verify, and sustain dependence on the literate for much information as well as for dealings with public and other agencies which require the use of written instruments. (Rahman 1993, 207)

This interdependence of literacy and power is important for it often engenders self-confidence, which is a cornerstone of people's development. However, the means of acquiring literacy also have a significant bearing on this. In spite of many efforts world-wide, including International Literacy Year 1990, which was proposed to UNESCO by the International Council for Adult Education (ICAE), and more than a decade of *Education for All* programmes, millions of adults in the Arab world, Asia, and Africa have still had little or no access to literacy learning opportunities. In fact, in a gender-segregated country such as Sudan, the majority of women have been left out of the education process. This is reflected in the high rate of illiteracy among Sudanese women, as in many other developing countries. For example, the illiteracy rate in Sudan is 72 percent but the gender data are 57.4 percent males and 83 percent females. There are two main reasons for this. First, in many rural areas, boys are sent to schools while girls are kept at home to help their mothers in the household, or because schools are outside the village. Second, school dropout rates among females are high due to early marriage. Third, adult women are burdened by household activities and have limited time to join literacy classes.

As a result, much more attention needs to be paid to developing nonformal adult learning processes, which take into account the high rates of illiteracy among girls and women, and work within an ecological context. But as Patrick Mucunguzi (1995, 341) notes "literacy is not a technique which can be deposited in the learners' minds as though the illiterate was an empty and ignorant container. Rather literacy teaching must begin with knowledge and the reality of the learners."

He goes on to say that literacy programmes in Uganda "have been characterized by teaching the people how to read and write without training them in the basic skills for the utilization of these resources" (Mucunguzi 1995, 34). Reorienting these programmes towards the needs of Sudanese women and the environment can increase their literacy levels and develop skills in efficient and sustainable agriculture. And, in fact, these are actually being done in Uganda.

The Multi-Purpose Training and Employment Association (MTEA) sponsors ecological literacy classes based on reading and writing the land. This new method or process of adult literacy is geared primarily towards women since it is women who more frequently enroll in literacy classes. The learning process has been weaving, described as a tapestry of food and learning, of colour, song, drumming, and theatre. "It is a blend of knowledge and the senses, rooted in the landscape and formed of local materials, serving the practical needs of the students yet assuring, by deft touches, its own self confident novelty" (Clover 1998, 1).

Through a process of adult education for ecological literacy, the women gain skills in marketing, organic, and traditional farming, and food production as they tend their own gardens. They also learn about marketing activities, re-learn how to identify, grow, and use many plants of the forests as traditional medicines. This is referred to by MTEA as "creating a knowledge forest" which means "a place where each plant is understood in terms of its importance to maintaining biodiversity, fixing nutrients into the soil, or as food, medicine, or building materials" (Clover 1998, 1).

The classes have been a major source of empowerment for many women. Not only have they learned how to increase production and sales of food, many have taken on leadership roles in their communities. Others have become civic educators around issues of sustainability, and/or mentors for other women. They have also gained the confidence and skills to design their own projects and learning, and developed strong opinions about the importance of education for their children. Creating literacy programmes within an environmental adult education context produces a more environmentally literate population and helps to change women's lives through its hands-on and relevant practices and approaches (Clover 1998).

Case Studies on Environmental Adult Education for Women

> Literacy programmes [must be] organised by people who care for the interests of all the citizens. It helps avoid the wasteful consumerist style of living perpetuated by the contemporary education system, which is still within the mainstream of the type of development characteristic of the colonial period. (Mucunguzi 1995, 341)

The Science Sector of the Sudanese National Commission for UNESCO is comprised of several different committees within UNESCO's fields of interest. Man [sic] and the Biosphere (MAB) is one of these. Environmental adult education is at the core of MAB activities to clarify and embrace public awareness on different environmental issues. These activities are targeted at different groups, with women occupying the central focus given their significant ties to the environment. The mass media are also very important, especially television and radio programmes on environmental and health education. One media project was organized by the Sudanese National Commission for UNESCO, in collaboration with the water and environmental sanitation (WES) programme of UNICEF, with the programmes broadcast by state television and radio which were supported by other national and regional institutions. For ten minutes a week, environmental and health education messages were broadcast across the country using dramas, interviews, and speakers.

The MAB committee also brings together governmental institutions, international agencies, and non-governmental organizations working in environmental and adult education, to develop coordinated activities. For example, World Environment Day in June 1996 included seminars and lectures for women and children of ways of protecting and conserving natural spaces and resources. Other activities have included long-term projects such as:

- tree planting in the nation's capital;
- discussions on environmental issues with residents in urban slums;
- information campaigns through the mass media;

- seminars on environmental health, industrial waste hazards and management, household garbage disposal, reforestation, the role of the media in environmental protection, and popular participation in environment improvement;
- a press conference on the National Environmental Plan of Action;
- seminars on the role of women in environmental conservation, protection, and health in cooperation with the Ministry of Health's School of Midwives, the Abu Halima Girls Training Centre, the Ministry of Social Welfare, and the Sudanese Family Planning Association.

While many MAB activities are centred in the nation's capital, special talks were also organized for female heads in the rural areas. Topics included issues of the environment from a global perspective, the role of women in environmental health, and the links between rural women, development, and environmental conservation.

In 1972, J. Roby Kidd, the founder of the International Council for Adult Education (ICAE), wrote that it was

essential [for the] mass media of communications [to] avoid contributing to the deterioration of the environment, [and], on the contrary, disseminate information of an educational nature on the need to protect and improve the environment in order to enable man [sic] to develop in every respect.

Thus, environmental adult educators should work to strengthen the role of the mass media by increasing their interest, effectiveness, and involvement in the development of television or radio broadcasts of environmental adult education programmes. Although not unproblematic, this effort is taking place in other parts of Africa. For example, in Botswana, radio programmes are "being used to disseminate to adult listeners, informational and educational messages about natural resource conservation and the protection and improvement of the environment"(Nyirenda 1995, 63). The content of these programmes is very diverse, coming from different "development areas or sectors such as urbanization, industry, health, agriculture and wildlife conservation" (Nyirenda 1995, 64). The programmes are intended to inform, motivate, and teach the general public. However, it is important to note that unless these programmes are aimed towards adults "participating in adult nonformal education activities or those who are in specifically organized nonformal environmental education activities such as study group campaigns on environmental conservation and protection," they will not be truly effective (Nyirenda 1995, 63–64). Open broadcasting is far less effective because an unorganized audience is "difficult to reach effectively, [and the] broadcaster is always uncertain that people are listening to the programmes" (Nyirenda 1995, 64).

Challenges Facing Environmental Adult Educators in Sudan

Non-participatory methods have been dominant in the execution of environmental education. (Mucunguzi 1995, 341)

Efforts to promote environmental adult education are increasingly evident among different institutions responsible for adult education. The National Council for Literacy and Adult Education (NCLAE) of the Ministry of Education takes a leading role in this. Nevertheless, there are a number of problems in Sudan with adult literacy work. For example, many NCLAE adult education programmes do not work effectively with all target groups. The problem may lie in the fact that NCLAE's work only focusses on the area of adult literacy where enrolment is significantly weak. Only a meagre one percent of the total population who are illiterate attend NCLAE's literacy classes. Even more problematic, perhaps, is the common tendency among many Arab countries, including Sudan, to associate adult education with literacy issues and not with its underlying goal of lifelong learning. Hence, the transformational value or role of environmental adult education as a catalyst to social and ecological change is overlooked.

Many other institutions carry out informal educational programmes that cover a broad range of environmental issues within the areas of agriculture, health, and nutrition, etc. Unfortunately, these programmes lack adequate continuity, consistency, planning, and direction, and the population they reach is small. Thus, they seem to have little significant impact on current environmental problems. The problem is compounded by the fact that NCLAE has been unable to coordinate the planning, implementation, and follow-up of these programmes according to its charter.

Given the high level of illiteracy in Sudan, as discussed earlier in this chapter, and the limited access to adult education, particularly for women, communications media such as television and radio could play a vital role in environmental adult education. However, this has not been the case and the media have failed to give due attention to environmental issues. This shortcoming may be attributed to the poor relationship between the media and the institutions of education. It can also be attributed to the fact that the communication systems in the less affluent Arab countries are generally poor and not accessible or available to all. This condition is further exacerbated by the chronic cuts and shortages in electricity, and the overall non-affordability of electricity.

Environmental adult educators in Sudan are working on a number of strategies to make their work more effective as a tool of socio-environmental transformation. One of these strategies focusses on empowering women and providing them with the skills and knowledge to make a stronger impact on the environment in policy and planning. This has meant increasing opportunities for the education and capacity-building of women as critical natural resource managers. and to impart these ecological values to their children.

A second area of work is the education of planners and decision makers on the connection between women and the environment, and how this relationship affects socio-economic development. Both women's knowledge and needs must be taken into account in any planning process if it is to truly be sustainable and worthwhile. In male-dominated cultures, women's knowledge and role is often not acknowledged or appreciated. Thus, these training programmes for planners and decision makers are of significant importance.

Environmental adult educators such as Kerrie Strathy (1995), Darlene Clover (1995), and Makkies David Lengwati (1995) have placed an emphasis on the potential of learning outdoors with women and girls. Nature cannot only teach about its problems but also can stimulate an emotional relationship and a call to action. This is especially important for urban dwellers who tend to have poor emotional relationships with the natural environment. Darlene Clover (1999, 235) argues that human beings are not just physically tied to the rest of nature, but are also emotionally, psychologically, spiritually, and culturally linked. She suggests that:

> Learning for ecological change should not and cannot be disconnected from the role of the rest of nature in the learning and teaching process. By learning and teaching in place we can be educated by the rest of nature.

Makkies David Lengwati (1995) notes that the rest of nature can provide lessons not only in its beauty and uniqueness, but also lessons on the consequences of environmental degradation, which are truly meaningful and engaging.

Conclusion

As environmental problems such as desertification continued to grow in Sudan, the quality of people's lives, and particularly those of women, continued to diminish. This prompted a movement away from *simply* developing an environmental education for children, towards a focus on adults. It was realized that there was and still is today, a critical need to develop environmental adult education strategies, particularly ones which are with, by, for, and about women.

In Sudan, environmental adult education is a process of working for socio-environmental change. It is about learning and building "creatively and sustainably for a healthier environment" (Lengwati 1995, 103). It is a process to better the quality of lives of all, but, in particular, to empower and enrich the lives of women who are most seriously affected, and who are also the most efficient and knowledgeable 'resource managers' in the country.

Environmental adult educators are in the process of developing training and capacity building programmes for women, influencing curriculum of the training of adults, and attempting to work more effectively with and through the mass media.

However, the high level of illiteracy in Sudan, particularly amongst women, is still a problem which needs to be addressed. Using literacy classes as a means to educate and empower women, as is happening in other parts of Africa, could be an important area for future explorations by environmental and adult educators in Sudan.

Note

1   For further reading in this area see Annabel Rodda (1992); Maria Mies and Vandana Shiva (1993); and Kerrie Strathy (1995).

References

Abdel Latif, E. 1992. "Women Contribution in Environmental Awareness Raising." Paper to Symposium on The Role of Women in Environment Conservation, Ministry of Social Welfare. Khartoum (Arabic).

Abu Affan, B.O. (Ed.) 1989. Proceedings of the Workshop on Policies and Strategies for Integrating Women in Agriculture and Rural Development. Khartoum.

Ahmed, A.J.M. 1991. Environmental Education: A Needed Change in Our Education: Khartoum: Institute of Environmental Studies, University of Khartoum.

Al Agib, Ibrahim. 1992. "Environmental Education in the Arab Countries." *Convergence XXV*, 2, 75–76.

Bhasin, Kamla. 1992. "Alternative and Sustainable Development." *Convergence XXV*, 2, 26–36.

Clover, Darlene. 1995. "Gender, Transformative Learning and Environmental Action." *Gender and Educatino 7*, 3, 243–258.

Clover, Darlene. 1998. "Ecological Literacy in Africa." *Pachamama, 3–04.*

El Jack, N.G. 1992. "The Impact of Drought and Desertification on Women." Paper to Symposium on The Role of Women on Environment Conservation, Ministry of Social Welfare, Khartoum.

El Nagar, S.H. 1991. "Women, Children and Environment in the Sudan." Paper presented to UNICEF, Khartoum.

Jong-Boon, C. (Ed.) 1990. "Women and Environment." Environmental Problems in Sudan. The Hague: ISS.

Khidir, M. and Mahmoud, M. 1991. "Women Issues in Development." Paper presented to the workshop on Integrating of Demographic Factors in Development Plans, Ministry of National Planning, Khartoum.

Kidd, Roby. 1972. "The Stockholm Conference." *Convergence V*, 3, 5–11.

Lengwati, Makkies David. 1995. "The Politics of Environmental Destruction and the Use of Nature as Teacher and Site of Learning." *Convergence XVIII*, 4, 99–105.

Mies, Maria and Shiva, Vandana. 1993. *Ecofeminism.* Halifax, Canada: Fernwood Publications.

Ministry of National Planning. 1991. "Report for the UN Environment Conference," Khartoum.

Mucunguzi, Patrick, 1995. "A Review of Nonformal Environmental Education in Uganda." *Environmental Education Research 1*, 3, 337–346.

Nyirenda, Juma E. 1995. "Radio Broadcasting for Adult Nonformal Environmental Education in Botswana." *Convergence XXVIII*, 4, 61–70.

Rahman, Anisur. 1993. *People's Self-Development, Perspectives on Participatory Action Research.* London: Zed Books.

Rich, Adrienne. 1989. *Time's Power.* New York: W.W. Norton and Company.

Rodda, Annabel. 1991. *Women and the Environment.* London: Zed Books.

Strathy, Kerrie. 1995. "Saving the Plants that Save Lives. SPACHEE/Fiji Department of Forestry Women and Forests Programme." *Convergence XXVII*, 4, 71–80.

Tabiedi, S.B. 1987. *Evaluation of the Role of Non-Governmental Organizations Providing Health Services in Sudan.* Unpublished M.Sc. thesis, University of Khartoum.

Chapter 6

Transforming Women's Lives
to "Save the Plants That Save Lives"
Through Environmental Education

Kerrie Strathy, Regina, Canada
Kesaia Tabunakawai, Suva, Fiji

Introduction

In the end we will conserve only what we love
We will only love what we understand
And we will only understand what we are taught
(Dioum, Senegal)

The Pacific Islands Developing Countries' (SPREP) report to the United Nations Conference on Environment and Development (UNCED) notes that the nations of the South Pacific region are custodians of a large portion of the Earth's surface. Their combined Exclusive Economic Zones (EEZ) occupy millions of square kilometres of the Pacific Ocean—an area three times larger than China or Canada and ten times the size of India. Their land area, however, is only 1.8 percent of that total and their total population is approximately 5.8 million.

Pacific Islanders depend on the biological resources of their small islands and the surrounding ocean to meet their needs. They have a close and special relationship with the environment because of their economic and cultural dependence on it. The main activities on the islands are fishing and agriculture and, for some islands, these are the only sources of export income. Island ecosystems are extremely fragile and must be handled with care if they are to continue to provide for the needs of current and future generations.

For many generations Pacific Islanders have used forest resources sustainably, and could continue to do so for future generations, if their living value is once again recognized. However, a changing world—which has put increased pressure on people to accumulate cash—threatens the survival of rainforests in Fiji and other countries in the Pacific. In the Solomon Islands, for example, it is estimated that, if current rates of logging continue, all useful timber species will be gone in less than seven years.

Since many of the islands that make up the South Pacific region are isolated from each other, they have a high degree of ecosystem and species diversity. Many of the forest plants found on these small land masses are endemic or found nowhere else in the world. The plants and forests on these islands are essential to the health and well-being of islanders who use them as food, fuel, and medicines, all of which are greatly needed in the region.

This chapter provides a case study of the development of WAINIMATE, the Women's Association for Natural Medicinal Therapy, which has its roots in a creative and innovative environmental adult education workshop entitled "Women and Forests." This project was organized by the South Pacific Association for Community Health and Environmental Education (SPACHEE), an adult education centre located at the University of the South Pacific, Fiji Campus. The workshop organizers took note of how women worked to meet both subsistence and cultural needs through forest resources, and, therefore, the ways in which they contribute greatly in social, environmental, and economic terms to the island's well-being. Given this, the goals of the workshop were to work towards the empowerment of women, and the transformation of their lives, and to 'save the plants that save lives.'

The Fijian Archipelago and the Issue of Health

Fiji is the third largest of the island nations of the Pacific and has more than 300 islands scattered over 1.3 million square kilometres of the South Pacific Ocean. The islands are characterized by diverse ecosystems, including extensive coral formations, mangrove, and forest areas that cover almost half of the land area. Although Fiji's vegetation is relatively small in number, it is of exceptional scientific and genetic interest given the high proportion of endemic species. Most of Fiji's 850,000 residents live on the largest island which has an area of 10,429 km^2, or more than half Fiji's total land area of 18,272 km^2 (Watling and Chape 1992).

The first European traders were attracted to Fiji in the early 1800s by the sandalwood trade. These traders made large profits but did no replanting, and therefore, by 1914, little sandalwood was left. When Fiji became a British colony in 1874, sugar cane plantations were established, with indentured labourers brought from India from 1879 to 1916. Their descendants now comprise approximately 45 percent of Fiji's population (Watling and Chape 1992).

Natural resource exploitation continues to support the economies of the South Pacific. Fiji's economy, for example, is dependent upon agriculture, forest, and fish exports. Its second largest industry (after sugar) is tourism, which is dependent upon the perceived image of pristine, white, sandy beaches. Continued exploitation of natural resources will ultimately lead to bankruptcy for Pacific Island nations as these resources are being harvested at rates which are not sustainable. Tourism will undoubtedly decline as the reefs and forests lose their spectacular range of biodiversity, and, consequently, their life (Watling and Chape 1992).

Fiji's Existing Environmental Problems

The tourist postcard image of Fiji and the rest of the South Pacific is generally one of a perfect paradise. But paradise, like the rest of the world, is facing a life-threatening assault on its environment. Some of the environmental issues currently facing the islands of the South Pacific include destruction of forests, mangroves, and coral reefs; declining fish stocks; contamination from pesticides and other toxic wastes; increasing volume and changing nature of rubbish; and rapid urbanization and population growth.

Many of the smaller islands and low-lying atolls in the Pacific are already being affected by the effects of climate change as sea-levels rise, supplies of freshwater become saline, and tropical storms increase in frequency and intensity. Natural disasters like cyclones, flooding, and drought are on the increase. Some argue that these potentially catastrophic events are, in part at least, a consequence of environmental degradation, and there is no doubt that they will lead to further environmental degradation (SPREP 1992).

Increasing and shifting populations are generating increasing volumes of rubbish that continue to be disposed of in mangrove areas, the sea, or rivers and ravines. While this practice was appropriate for the organic rubbish produced in the past—coconut husks, vegetable peelings, and seashells—it is inappropriate for materials such as plastics, batteries, and bottles which are now being discarded. Industrial and tourist development along coastal areas leads to waste being dumped into the sea or rivers. Some environmental problems are the result of over-exploitation of natural resources and/or inappropriate harvesting techniques. Fish poisons continue to be used in some areas to harvest fish, and consequently many small fish and corals are also being destroyed in the process. Pesticides and dynamite have replaced the old plant-based poisons in some areas, with quite serious and far-reaching consequences (SPREP 1992).

There are many areas where farming is carried out on steep slopes without any mitigating factors such as terracing or agro-forestry, which would reduce soil erosion. In many cases, these areas were cleared of forests to make way for food crops and cash crops, such as ginger. Commercial logging has also been carried out on inappropriate areas, such as along steep slopes. This is happening in spite of a Fiji Code of Logging Practice, which was developed to confine logging to areas that are not so environmentally vulnerable.

The loss of traditional knowledge is a serious problem, particularly around traditional medicines in the South Pacific. Many people in Fiji have limited access to modern health services, and even fewer have access to modern medicines. One doctor recounted a story of how he was forced to use a traditional medicine on leg ulcers when rough seas delayed delivery of medical supplies to a remote island for several weeks. He had learned about this medicine from his mother who was a traditional healer, and found that it actually worked better than penicillin. Another doctor told of how he made a medicine of beach morning glories to treat the eye infection of a young girl who was not responding to treatments prescribed by numerous doctors. Many nurses also admitted to administering traditional medicines that they knew were safe and effective to patients in hospital, but frequently these are given at night when doctors are not around. When that traditional knowledge is lost, when the forest resources are gone, villagers will become dependent on non-traditional medicines, instead of using the plants that they can obtain easily and for free from their forest (Thaman 1992).

Often, due to their different roles in society, men and women have different knowledge about plant uses (Sontheimer 1991). The Community-Based Biodiversity Conservation (CBBC) Project in Fiji found that women were frequently more knowledgeable about the properties and use of plants than men were, even in the same village. This makes sense, given their multiple family and community responsibilities. Some men even claimed that terrestrial plants were found growing in water. A study carried out in Ghana found similar results—women identified thirty-one uses of trees on fallow land, while men could identify only eight (Marstrand 1991). The challenge, however, is for women to speak with confidence about local development decisions—largely made by men—which could have a detrimental effect on the women's continued ability to access plant resources, such as those used in making traditional medicines.

Why Environmental Education is Important for Adults

According to Fiji's National Environment Strategy, environmental education is slowly being integrated into the school curriculum but educational processes that respect, engage, and motivate the wider community need special attention. The strategy also suggests that developing this type of education is likely to be more effective and efficiently undertaken by non-governmental organizations (NGOs), than by government. As a result, it urges government to facilitate the operation of NGOs. In light of the increasing pressure to exploit forest reserves for short-term cash, it is imperative that educational programmes be developed for adults as a means of encouraging action towards better conservation.

Since many of the plants found in the South Pacific are used primarily by women, the SPREP report (1992) to UNCED highlighted the need "to facilitate the access to environmental information for all groups, in particular women and youth, in order to enhance management of resources and the environment." Chapter 8 of *Agenda 21* of the United Nations Development Fund for Women (UNIFEM) also recognized the need to involve all people concerned in decision making and land use management. The need to involve women in managing fragile ecosystems was highlighted in Chapter 12, while the particular role of women in the conservation of biological diversity was emphasized in Chapter 15.

In response to concerns of local women's organizations, and the above considerations brought back to Fiji by the SPACHEE representative to United Nations Conference on Environment and Development (UNCED), a pilot 'Women and Forests Workshop' was organized. Women living in the interior of the larger Pacific islands, such as Fiji, depend on the forests to meet their basic needs. Forests provide Pacific women and their families with food, fuel, medicines, craft materials, and more. Workshop organizers recognized that women will continue to be able to meet the subsistence and cultural needs of their families, and to contribute to its economic needs, if they are able to come together and strategize ways in which they can use the forests wisely, and challenge policies which counter-act that type of usage.

WAINIMATE's Environmental Adult Education Philosophy and Framework

The pilot "Women and Forests Workshop," and subsequent traditional medicine workshops organized by WAINIMATE, were based on fundamental principles of adult education, which have been articulated by numerous adult educators worldwide. The first principle, based on respect for the adult learner, meant that workshop facilitators viewed the participants as knowledgeable and, in fact, as the 'experts.' This meant that resource persons were kept to a minimum, and organizers functioned as facilitators to encourage the women in these workshops to share their knowledge about the traditional uses of forest plants. They were sensitive to the different levels of understanding within the group, and endeavoured to provide a forum where all felt comfortable to intervene and clarify any issues unfamiliar to them.

The workshops always start from the women's own experiences and are based on dialogue and creativity to stimulate interest in the workshop, itself, and afterwards. The programmes are also designed to be action-oriented, with the women being encouraged to take action to protect forest ecosystems in general, and to work towards the continued availability of medicinal plants in particular.

Those selected to be at the first workshop were primarily leaders of women's organizations. This was the case because a major aim of the workshops was to encourage participants to share their experience with other women, in an effort to stimulate them to protect forest ecosystems and medicinal plants. Essentially, the workshop experience sought to empower women to take collective action to protect the forest resources they are dependent upon. This was thought to be most likely to happen with women who were already involved in social movements of one kind or another.

A third set of educational principles that helped direct the learning approach are those generally ascribed to feminist adult education. The organizers recognized that women have different experiences and, consequently, different knowledge which must be taken into account in order to bring about socio-environmental change (Clover et al. 2000). It was also recognized that women are not normally involved in community decision making in most parts of the Pacific—even when decisions have a direct bearing on their own practical needs and strategic interests. The workshop sought to encourage women to recognize their knowledge and themselves as 'experts,' to be more assertive, and to use the communication skills they were developing.

Throughout history women have initiated certain systems of knowledge. For example, they were the first healers in society, the first medicine women. However, gradually, when creation of knowledge became an institution, when it became formal, men took it over almost completely. Men, especially upper caste and upper class men, have been creating knowledge, and have controlled this knowledge, to oppress others, to disempower them. Male historians write history, and leave *her* story out (quoted in *People's ACCESS* 1991, 55).

Formal education systems, to a large extent, have also dismissed experiential knowledge. This is particularly evident with women who have practiced traditional medicine. They were the women burned alive as witches in Europe. These knowledgeable women—medicine women and midwives—were a threat to the male hierarchy and the religious leaders who eliminated them. The experiential knowledge held by traditional healers in the Pacific and elsewhere is currently under threat from western-trained medical professionals who are similarly trying to maintain positions of power and control. It is not uncommon to hear traditional medicine practitioners referred to as 'quacks' or 'witches,' and there is still legislation in the Pacific prohibiting the practice of traditional medicine, even though there is much evidence to justify these practices and practitioners. However, our approach was also based on the concept of Primary Environment Care (PEC), whereby local skills and knowledge are used to care for natural resources while satisfying livelihood needs, such as the provision of medicinal plants (Pretty and Gujit 1992).

The workshop relied on environment adult education practices of using the rest of nature as a teacher and site of learning. Gordon (quoted in George and Griffen 1992, 22) argues that the forests are where we can best teach our children "about the old ways, their responsibilities to their family and community, and their obligations to the land." We need to experience the relationship of plants within the forest ecosystem—how the creepers and vines depend on the trees for their nourishment—the way a baby depends on its mother for milk.

Women and Forests: The Practice

The organizers of the 1992 "Women and Forests Workshop" chose not to hold the weeklong workshop entirely in Suva, Fiji's capital. Instead of hearing about forest ecosystems and the need to conserve biological diversity, participants spent five days learning with Nadovu villagers who live in, and depend upon, Fiji's productive rainforest. This approach had very positive benefits for the women of Nadovu, for the community as a whole, and for the workshop participants who came to know, understand, and appreciate the value of rainforests.

The workshop began by asking everyone to name her favourite tree and explain the reason(s). Many participants named the coconut tree because all of its parts (leaves, husk, fruit, wood, etc.) are useful. They also identified fruit and nut trees, trees used to supply craft materials, trees used for traditional medicine, and a variety of others such as hibiscus, which also have multiple uses. This initial exercise seemed to clearly demonstrate the importance of biological diversity to women, their families, and their communities.

The women then looked at various ways to communicate messages about environmental protection and forest conservation during the workshop. They discussed the importance of creativity in the learning process and how songs, posters, and videos can convey important information and feelings. Songs, such as *Take Care of Our Island* and *Mati Sobu, ua mai* (High Tide, Low Tide) were so popular that participants wanted to perform them at the public opening that was held on the first evening. They later wrote their own songs and poetry. In fact, plans are underway to record an entire tape of environmental songs created at a number of workshops, and to use the funds they raise to generate future programmes. Moreover, it has been suggested that tee shirts and/or *sulus* (wraps) could be printed with clear messages as a means of stimulating environment awareness throughout Fiji and other parts of the South Pacific, to serve the same purpose.

The chief guest at the official opening of the workshop was a well-known Fijian radio personality and traditional healer, Sokoveti Ravovo, who made a very stimulating presentation. This resulted in a lengthy discussion that continued throughout the workshop. It was, therefore, not surprising that plants and traditional medicine became the focus of future women and forest programmes.

During the first workshop, the women had an opportunity to see the devastation that could be caused by a small logging operation. This opportunity came from the manager of the sawmill in the next village who heard we were in the area and invited the participants to come and see it. However, little did he know that the focus of our workshop was on forest conservation! The women saw leftover bits of trees pushed down hillsides and sawdust lying in heaps that were then washed towards the river—along with the soil that was being eroded as a result of the loss of forest cover. (This helped explain why the river flowing through Nadovu, where people bathed and washed their clothes, was red and muddy after torrential rains the day before.) The sawmill itself was leaking oil, which was also running towards the river.

The visit to the sawmill had a major impact on the women, many of whom have undertaken follow-up activities since, including ongoing community learning programmes related to forest conservation and protection, planting traditional *kau salusalu* plants, fruit, and medicinal trees in their own villages. A 'Rainforest News Alert' education kit was prepared and a video, *Protect the Earth and She Will Provide for You* produced. One participant carried out research into pulp logging operations in Fiji, and others developed the Lali Theatre Company, as a result of the enthusiastic response they had to the popular theatre video they saw. The Lali Theatre had its first performance at a public event to mark World Environment Day in June 1993. Although the performance was very well received, the group members were unable to continue due to other commitments.

Since many of the women at the workshop had a particular interest in traditional medicine, they played an active role in the planning and implementation of the First Regional Women's Traditional Medicine Workshop, which was held in August 1993. They wanted to encourage the sharing of information about traditional medicines, and the threats to them from rainforest destruction, with their sisters in the Pacific. This workshop, which was jointly organized by SPACHEE, Fiji Forestry Department, and the Young Women's Christian Association (YWCA), generated so much interest that the University of the South Pacific (USP) and the South Pacific Forestry Development Programme (SPFDP) offered assistance. Assistance was also received from the German Technical Cooperation Agency's (GTZ) Fiji-German Forestry Project, the International Women's Development Agency (IWDA), and New Zealand's Official Development Assistance agency (ODA), for the Second Regional Traditional Medicine Workshop, held two years later.

The aim of the workshop was to use a creative and innovative environmental adult education process to encourage the documentation and promotion of indigenous knowledge about medicinal plants. The Fijian participants brought samples of the plants used for traditional medicines while the regional participants brought photographs of their medicinal plants. The workshop actually started the day before the official opening when participants identified a medicinal use for virtually every plant they encountered on the short walk from the dormitories to the dining hall! There was much excitement when they learned that women from other islands were using the same plant species, or that plants they were not currently using had medicinal uses in other countries.

SPACHEE's chairperson, Dr. William Aalbersberg, discussed with participants the need to document traditional medicine practice. He also outlined some research initiatives USP is involved in to ensure the efficacy and safety of medicinal plants. The issue of indigenous knowledge and the need for researchers to recognize and respect this knowledge, as well as the rights of landowners to compensation for their knowledge and use of their plant species, were also raised. Dr. Ken Chen, a World Health Organization

(WHO) regional adviser for traditional medicine, spoke about WHO's interest in traditional medicine and stressed that traditional medicine has an important role to play in achieving WHO's goal of *Health for All by the Year 2000*. Dr. Wolf Forstreuter, Fiji Forest Inventory Project, explained that minor forest products, such as traditional medicines, are included in the national forest inventory, in recognition of the fact that a forest provides much more than timber.

The rest of the programme, like the earlier one, involved participants as resource persons and facilitators of discussion groups on the issues to be addressed. A key concern was that some traditional medicine plant species were becoming difficult to locate. Another issue was the proper identification of medicinal plant species, and the conditions they were used to treat. Cook Island participants reported on the traditional medicine demonstration garden that was started by the Conservation Department, while participants from Tahiti talked about the Traditional Healers' Association they started in the early 1980s.

On a one-day field visit to the Nakavu Natural Forest Management Project (of the Fiji-German Forestry Project - GTZ), an inventory of traditional medicine plants was carried out. The participants developed follow-up plans to encourage the promotion of traditional medicine practice and the conservation of traditional medicine plant species. These plans included establishing national Traditional Medicine Associations and demonstration gardens, similar to the one they established at USP as part of the official closing ceremony. Representatives from the Fiji School of Medicine, the World Health Organization, United Nations Development Programme (UNDP), Food and Agriculture Organization of the United Nations (FAO), the World YWCA, local embassies, and women's organizations planted trees in the demonstration garden. Workshop participants also adopted the Pacifica Declaration on Traditional Medicine.

The Fijian participants went on to organize follow-up workshops at the divisional level, followed by a national workshop to review progress in the campaign to 'Save the Plants That Save Lives.' Assistance from the Canada Fund enabled Fijian healers to establish a series of local traditional medicine demonstrations gardens throughout Fiji, and to work on the Fijian Traditional Medicine Handbook, conceived at the August 1993 workshop. This handbook—or 'cookbook' of traditional medicines—was prepared for village women who do not have access to other health services, or the cash to purchase medicines for basic first aid treatments. They have also helped verify the documentation compiled for the WHO publication on traditional medicine in the South Pacific by chemistry professors at the University of the South Pacific.

At the closing ceremony for the Second Regional Women's Traditional Medicine Workshop—held on World Health Day (April 7th)—Fiji healers and conservationists were pleased to announce that they had worked together to register WAINIMATE (Women's Association for Natural Medicinal Therapy) as a charitable trust. These healers who shared their experience with those at the workshop are anxious to encourage their Pacific sisters to establish their own organizations, to ensure the conservation of traditional medicine plants, and the integration of traditional medicines into national health delivery systems throughout the Pacific.

WAINIMATE has continued to work with SPACHEE, the Fiji Forestry Department, and others to encourage the conservation of medicinal plants, since the Second Regional Workshop. WAINIMATE has carried out a series of Divisional workshops in partnership with the Asia South Pacific Bureau for Adult Education (ASPBAE). In addition, a series of workshops was organized around the issue of medicine-making, with the support of the World Wildlife Fund's (WWF) South Pacific Programme. Importantly, the workshops were organized largely at the request, and with the assistance of, the women who had taken part in the earlier workshops.

In an effort to increase awareness about the work of WAINIMATE and the need to 'save the plants that save lives,' WAINIMATE members organized the Third Regional Women's Traditional Medicine Workshop, which used the slogan 'Educate a Woman, Educate a Nation.' This workshop brought together more women from other Pacific Island countries, and was timed to coincide with the Pacific Science Congress held at USP in July 1997. WAINIMATE members and workshop participants mounted very impressive displays and demonstrations as part of the ECOWoman Pavilion, set up to showcase women and environment activities at the Science Congress. Hundreds of visitors, including many high school students, accompanied traditional healers on tours through the USP campus and botanical garden to learn the names and uses of medicinal plants.

The Transformative Potential

The forest and traditional medicine workshop programme has proven to be a transformative learning experience, in that a number of traditional medicine practitioners and workshop participants have radically altered their thinking and their practice. One participant from the First Regional Women's Traditional Medicine Workshop, for example, now introduces herself as a community worker, rather than as a housewife. She has become very involved in her traditional medicine practice, and in convincing others to be concerned about the potential loss of medicinal plant resources. At the Pacific Science Congress, WAINIMATE members pointed out to Department of Agriculture staff that the containers of weeds they had were, in fact, useful traditional medicine plants!

Other traditional healers have also come to value the knowledge they have accumulated over the years. One healer, who took part in the traditional medicine conservation workshop held in Savusavu, thanked the organizers for helping her to recognize the value of her knowledge of traditional medicine. Prior to the workshop, she did not consider that what she knew was important. But the discussions, field visit, and subsequent medicine-making demonstrations helped her to see otherwise. Many healers have expressed similar sentiments at other workshops and demonstrations.

The issue of intellectual property rights was raised at the First Regional Workshop and discussed further at the second. Some of the healers recounted stories of foreign researchers who took samples of the medicinal plants they use. Work has subsequently been carried out to develop laws in Fiji to protect indigenous plants and plant knowledge from foreign exploitation. Previously, nothing existed to safeguard these rights, and nothing would have changed if indigenous people had not organized themselves to defend their interests through organizations such as WAINIMATE. We are hopeful that the Fiji experience can help to stimulate and mobilize other Pacific nations to deal with these issues as well.

Participants at all workshops have been enthusiastic and keen to initiate follow-up activities at home, in other parts of Fiji, and in other participating countries. The video on traditional medicine, *Grandma's Cure,* produced after the First Regional Workshop, has been shown to hundreds of women and men throughout Fiji and other parts of the Pacific, in an effort to involve more women in the campaign to 'Save the Plants That Save Lives.' In Kiribati, another association of traditional healers was formed as a result of the Second Regional Workshop. Follow-up visits by WAINIMATE members to Tonga, Nauru, Solomon Islands, and Vanuatu led to the establishment of local traditional medicine groups. It is hoped that groups will soon be formed in other Pacific Island countries.

Increasingly, men are taking an interest in our workshops, even though most of the healers in the Pacific are women. Men are anxious to know what their mothers and wives are learning, and are often found with their *yaqona* bowl in a corner of the meeting room, or outside by an open window or door. *Yaqona*, or *kava* as it is more popularly known, is a drink made from the pounded root of the *piper methsticum*, a common garden shrub, and has played a very important role in traditional Fijian ceremonies for generations. Women in Ucunivanua and Vagadaci were able to get support from the village men to set aside a medicinal plant reserve, after a workshop in their village. It is envisioned that this practice will be a continuing feature at future workshops.

A presentation on the First Regional Workshop, made to the predominantly male South Pacific Heads of Forestry meeting in 1993, generated a great deal of interest and discussion. As a result, the final statement from the meeting included reference to the medicinal value of forests and the need to conserve medicinal plants. The Heads of Forestry also identified the need to carry out research on the usefulness and income-generating potential of medicinal plants that could allow communities to retain and use their forest resources more sustainably. Their support encouraged the South Pacific Forestry Development Program—later renamed the Pacific Islands Forest and Trees Support Program—to continue providing support to WAINIMATE.

These transformative learning experiences continue to empower women involved with WAINIMATE. Several members of WAINIMATE have been able to take advantage of opportunities to learn alongside men in forestry workshops, and to share their vast wealth of knowledge about medicinal plants with professional foresters. The experiences provided through WAINIMATE workshops gave traditional healers the confidence to work with these foresters, to play an active role in the development of Fiji's Biodiversity Conservation Strategy, and to ensure the Strategy included provisions to safeguard the biodiversity interests of traditional healers and other women.

Fiji's Ministry of Health has also taken an interest in WAINIMATE's work. Although it will take some time for the ministry to fully recognize the value of traditional medicines, and the healers who prepare them, the channels of communication are open. Discussions are being held about the integration of traditional medicines into the overall health delivery system, as a means of reducing dependence on increasingly expensive imported western medicines. This is significant progress when one considers that our initial meeting with Ministry of Health staff began with skepticism about the contribution traditional medicine could make to modern health delivery.

We used World Health Organization declarations as a starting point in our discussions with the Ministry of Health. This led WHO to explicitly express its support for WAINIMATE and to request that the Ministry of Health provide assistance for our workshops and the handbook on traditional medicine compiled by members. The handbook was officially launched at the World Health Day event, organized to celebrate WHO's 50th anniversary in 1998—a significant accomplishment. As a result of this very empowering opportunity, WAINIMATE members became even more determined to convince the Ministry of Health to recognize the valuable contribution safe and effective traditional medicine could make to achieving WHO's goal of 'Health for All by the Year 2000'—and to make it affordable.

The Rarotonga Agreement Towards Healthy Islands, adopted by Ministers of Pacific Island countries at their meeting held in August 1997, included a section on traditional medicine, in which WHO committed itself to support national level workshops on traditional medicine and "to encourage incorporation of traditional medical practices into health systems in the Region." WAINIMATE members were subsequently asked to organize and facilitate a two-day workshop on traditional medicine for approximately eighty of the Fiji Ministry of Health's top medical staff, including divisional medical officers and nursing disters, in late 1998. The fact that all but one of the workshop participants admitted to using traditional medicine, or to referring a patient to a traditional healer, was another significant milestone along our path towards the recognition of traditional medicine and traditional medicine practitioners. Less than a year after this workshop the Ministry of Health provided WAINIMATE with an office and meeting space.

Our next challenge is to have the ministry recognize traditional healers. This will undoubtedly be more difficult. But as more women are involved in our efforts we are finding more support amongst western trained medical personnel, who can testify to the safety and efficacy of medicines administered by traditional healers. Some doctors claim that certain conditions respond better to traditional medicine and advise patients to use traditional medicines, or to seek the assistance of a traditional medicine practitioner. It is not uncommon for doctors to recount stories of patients whose conditions have cleared up after using traditional medicines, many of whom only sought the assistance of traditional practitioners when their conditions did not respond to conventional western medicines.

Conclusion

The women and forests activities carried out by WAINIMATE, with support from the Fiji Forestry Department and many others, have been very well received by women in Fiji and throughout the Pacific. Workshop participants have enthusiastically planned follow-up workshops and look forward to the opportunity to share their stories. They are interested in the traditional medicine conservation programme because it provides them with practical learning opportunities that are relevant to their concerns and interests.

It is important in the learning process that participants have fun by sharing in laughter (Pretty 1995). This was certainly true for the Women and Forests Workshop, and the subsequent traditional medicine conservation workshops. The women who participated in a *Taveuni* workshop said that they thoroughly enjoyed themselves because they were able to express themselves, and give their ideas and views, too. Through the campaign to "Save the Plants That Save Lives," women in the Pacific are learning more about the need to protect their environment as well as how to better address the primary health care needs of their families and communities, in an atmosphere that is both fun and empowering.

The women who formed WAINIMATE represent one example of how Pacific women are becoming mobilized to do this. A sustainable future for fragile island ecosystems, and the Earth, depends upon a healthy environment, economy, and population. The future well-being of Pacific Islanders is dependent upon using women's knowledge and commitment to protect fragile island ecosystems. WAINIMATE members know that our own survival, and the survival of future generations, is seriously threatened as the area of forest cover is reduced. These actions jeopardize the continued availability of medicinal plants and other life-giving resources found in the forest.

Women in the Pacific—like women around the world—play a very important role as educators, communicators, and activists within their family, workplace, and community. Their motivation to take care of their islands will be based on the recognition that, without a healthy environment, there is no healthy life. Women will, undoubtedly, continue to be central players in the move towards sustainability, given their close relationship with nature and concern for the future well-being of their children, families, and friends.

Practices to Ensure Survival of Medicinal Plant Resources

1 Respect biodiversity—save one, save all.
2 Establish medicinal plant reserves.
3 Grow your own medicine at home, health centre, or village dispensary.
4 Use mature trees.
5 Pick only what you need.
6 Use bark on one side only.

7   Leave the main root behind.
8   Be careful with the blade (grass cutter).
9   Preserve the seeds for future generations.
10   Cut one, plant more!

Our Life Depends on Plants—If They Die We Die!

References

Clover, Darlene E., Follen, Shirley, and Hall, Budd L. 2000. *The Nature of Transformation: Environmental Adult Education.* Toronto: LEAP and OISE, University of Toronto.

Dioum, Baba. 1992. *Dictionary of Environmental Quotations.* Barbara K. Rodes and Rice Odell (Eds.). New York: Simon and Schuster.

George, Susanna and Griffen, Vanessa. 1992. "Asian and Pacific Women's Resource Series." Kuala Lumpur: Environment, Asian and Pacific Development Centre.

Marstrand, Pauline et al. 1991. *Sustainable Development: An Imperative for Environmental Protection.* London: Commonwealth Secretariat.

People's ACCESS. 1991. "Adult Education in the 90s: Unity in Diversity." Proceedings from the First ASPBAE General Assembly on Adult Education. Tagaytay City, Philippines, December 8–14.

Pretty, Jules, Guijt, Irene, Thompson, John, and Scoones, Ian. 1995. *Participatory Learning and Action: A Trainer's Guide.* London: International Institute for Environment and Development.

Pretty, Jules and Gujit, Irene. 1992. "Primary Environmental Care: An Alternative Paradigm for Development Assistance." *Environment and Urbanisation 4, 1.*

Sontheimer, Sally. (Ed.) 1991. *Women and the Environment: A Reader Crisis and Development in the Third World.* London: Earthscan Publications.

SPREP. 1992. "The Pacific Way: Pacific Islands Developing Countries' Report to the United Nations Conference on Environment and Development." Noumea: South Pacific Commission.

Thaman, Randolph. 1992. "Environment Awareness and Conservation." Unpublished paper presented at SPACHEE Women and Forests Workshop, Suva, July.

UNIFEM. 1993. "Agenda 21: An Easy Reference to the Specific Recommendations on Women." New York: UNIFEM.

World Health Organization. 1997. "The Rarotonga Agreement Towards Healthy Islands." Unpublished Report of the Meeting of Ministers of Health for the Pacific Island Countries. Rarotonga, Cook Island: August 6–7.

World Health Organization. 1998. *Medicinal Plants in the South Pacific.* Manila: WHO.

Watling, Dick and Chape, Stuart. 1992. "Fiji State of the Environment Report." Suva, Fiji: Government Printer.

Chapter 7

Environmental Popular Education Strategies for Creating Sustainable Communities in the Pátzcuaro Lake Basin Region of Mexico

Joaquín Esteva and Javier Reyes
Mexico

The ecology and productivity of the Pátzcuaro Michoacan Lake region of western Mexico has been endangered as a result of modernization processes that began in the last century, but have accelerated in recent decades. The major problems are soil erosion, caused by deforestation for lumber and farming, and pollution from urban sewage. Soil erosion continually obstructs the water conduits of the lake, and sewage dumping increases the eutrophication of the water, placing the lake at great risk of premature death (Chacón 1994; Barrera 1993).

In response to these growing problems, and with the purpose of creating a more equitable, democratic, productive, and sustainable regional society, the Centre for Social and Ecological Studies (CESE) began in 1983 to facilitate environmental popular education activities in the Pátzcuaro Michoacan region. One major understanding to which CESE came through this work was that gender relations are a critical element. Without acknowledging gender inequities and women's needs and wisdom, ensuring that they have opportunities to share their knowledge, and providing the space for their full participation at all levels of change and decision making, a sustainable society will never be reached.

This chapter examines the work of CESE through its conceptual framework, guiding principles, objectives, and educational activities. To demonstrate the work and principles, particularly those addressing unequal gender relations, a case study of one project targeted specifically at women is provided.

Socio-Environmental Profile of the Pátzcuaro Region

The Pátzcuaro Michoacan Region of Mexico is home to both urban and rural Indian communities. The people belong to the Purepécha ethnic group. The total population of the lake basin is 120,000, distributed over an area of 110 km^2. The area has a temperate climate and a varied ecosystem that includes pinetholm oak forests, volcanic lava regions, inter-mountainous valleys, and the Pátzcuaro Michoacan Lake (Barrera-Bassols 1993).

Traditionally, communities have made rational use of resources to meet the basic needs of the local population and to sustain the environment. However, this has changed with the introduction of new techniques and modern practices in the most important areas of production: agriculture, fishing, and crafts.

Today, the social dimension in the Pátzcuaro region is characterized by the presence of both traditional practices and modern techniques that have not entirely displaced traditional practices. Unfortunately, the interaction between the traditional and modern methods of agriculture and fishing which began in the 1940s has not been properly organized and is, therefore, failing to meet both the needs of the rural population and the need to protect the local environment (Esteva 1995).

Politically, the State of Michoacan has transformed the lake region into an 'Indian' showpiece for the social politics of successive state governments. This has resulted in a tightening of state control over budgets and environmental programmes, which has marginalized local communities and governments and excluded them from many important decisions. Thus, the people of the lake region are unable to participate in democratic processes that seek solutions to local environmental problems. However, in spite of this, the citizens have made some political advances themselves in relation to the protection of the region's environment.

Conceptual Framework

The fundamentals of CESE's work revolve around the areas of sustainability, environmental popular education, and participatory research.

Sustainable Development

We find general references to sustainable development in environmental politics. Accordingly, the present day planetary crisis should be addressed with changes in three relationships. The first of these involves the present social relationships and lifestyles engendered by neo-liberalism that promote poverty, injustice, and unequal distribution of power. Another is the relationship between society and the environment that has been primarily one of ecocide, resulting from the rise of urbanization and industrialization that perpetuates a predatory, consumerist development. The final relationship is that between people and their social and natural environments. The change should be based on the development of a new ethic in regards to both human and non-human life, and the development of values and practices that assure both material and spiritual well-being for each individual.

From the social perspective, a gender component that involves an analysis of the transformation of relationships between men and women, and in turn their relationship with nature, is being incorporated into our vision of sustainable development.

From CESE's perspective, sustainable development implies the affirmation of an indissoluble relationship between development and environment that encompasses, among other things, the possibility that every individual and community can live satisfactorily when ecological balance is understood and respected. Included in this is the recognition that structural strategies need to be introduced to combat poverty by transforming the current patterns of production, consumption, and resource distribution, and to ensure that the necessary resources are available to meet the needs of future generations.

The transition towards a sustainable society cannot be achieved without an informed population and the democratic processes necessary to avert the negative impact of neo-liberalist policies of governments and multinational corporations, both locally and globally. Nevertheless, achieving these aims should not represent an end in itself, but a way to approach a sustainable society. Mere political change will not achieve anything on its own, as is proven by the experience of socialist countries that have followed the urban-industrial model of development in capitalist countries while forming a new society. Those results have been disastrous, both socially and ecologically.

Latin American countries are considered to have among the richest biodiversity on the planet, and thus these countries are confronted with the huge responsibility of safeguarding this enormous, yet fragile ecosystem. As many traditional cultures promote respect for the environment, the acceptance of cultural diversity is an essential factor in moving towards a sustainable society.

Environmental Popular Education

The challenges that confront contemporary Latin America require popular educators to reflect not only on their everyday practice, but also on their perception of the *raison d'etre* for education. The current spiritual, material, and ecological crisis suggests the need for a paradigmatic review of popular education. Finding solutions to these crises will require the repudiation of the anthropocentrism that still characterizes education in general.

Current planetary conditions necessitate that people undertake enormous theoretical, practical, and financial efforts to advance a new type of development. There are individuals in society who dedicate considerable energy to the formulation of international, social, and environmental policies through numerous networks and organizations, locally and abroad. Locally, this strategy points to the strengthening of social groups in the natural cycle of action and decision making. It is here that environmental popular educators still have a particularly significant role, not only in the multiple functions they have traditionally fulfilled, but, more importantly, in understanding new paradigms that have emerged out of the increasing integration of social and natural disciplines.

For many this environmental link is useful because it offers elements and dimensions within which a critical analysis of society can be made. At the same time, it permits the curricular organization of subjects with a new integrating axis, the environment, and to have ecosystems as line didactic instruments. The instruments of observation, registration, experimentation, systematization, analysis, and interpretation of results are enriched with the addition of disciplines such as ethnobiology. Finally, in ecology, a social binding factor can be found that allows the teacher to collaborate in a structured process for organized social participation.

The task for environmental popular education should be *concientizacion*, that is, drawing out an embedded environmental consciousness through the collective development and acquisition of new knowledge for the development of a more sustainable society. This implies the development of educational methods and strategies that are not neutral but rather help to create new ethics for human and non-human life, to form equitable relationships between society and nature, and to introduce processes of information and reflection that lead to the re-appropriation of nature and society, both conceptually and practically.

In this sense, the pedagogical process cannot restrict itself to the introduction of ecological information to environmental popular education programmes. This conservative strategy is flawed in that it fails to recognize other perspectives about the environmental crisis that we face. Instead, environmental popular education should be a critical and creative process based on the idea of *concientizacion,* which respects knowledge and engages people in critical dialogue to create new sets of knowledge and learning paradigms. These are oriented by a principle of reality feedback that tends to conceive of it as a complex interdisciplinary system among physical, biological, and social processes (Leff 1992).

Environmental popular education is an ethnic-political-pedagogical option that supports the environmental movement. It is defined as a lifelong learning process from a political perspective. In the construction of sustainable societies that respond to existing cultural and ecological particularities, grassroots environmental education encompasses theoretical and practical elements that are aimed at modifying attitudes, raising levels of understanding, and enriching people's lives within the context of their socio-cultural relationships, and their biophysical medium (Reyes 1994).

Participatory Research

Since the 1970s, participatory research has been an important addition to the Latin American movement in the development of new methods of social research. Participatory research is a social process of knowledge production. It is also an educational method and an instrument for consciousness-raising. Its objective is to understand and analyze a reality in its three constituent moments: the objective processes; the perception (level of consciousness) of these processes in a person; and the lived experience within a concrete structure. It is characterized as a qualitative research methodology in which quantitative elements can be contemplated, but always within the context of qualitative questions. When compared with other forms of social research, participatory research offers a process that is more in line with the principles of adult education, more immersed in action, and more scientifically oriented towards producing a complex and true image of reality (for examples, see Bosco 1977; De Schutter 1996; and Hall and Gillette 1977).

Research that is related to aspects of social life needs to involve people in a different relationship from research, which sees subjects as objects. Technically, participatory research can be defined as:

> the process of production and appropriation of knowledge, decisions and actions that tend to change social relationships based on a dialogue between researchers and social groups as well as relating the research to communication, organization and training. (CREFAL 1980, 83)

Participatory research, like popular education, has as its main point of reference a paradigm that is centred on the knowledge and the transformation of social relationships. By assuming an environmental paradigm as a point of departure, participatory research has added methods that favour the conjunction of scientific knowledge and popular environmentalism in the building of a sustainable society.

Presently, community social-environmental research is the expression of a new synthesis between natural and social-scientific knowledge as is the case in the research and planning for territorial organization or integrated management of natural resources with participatory methods. These efforts of research and planning form part of training programmes for the sustainable management of natural resources, which is a priority in contemporary Latin American environmental educational strategies (Network for Popular Education and Ecology—REPEC 1996).

Primary Objectives and Projects of CESE

Operating within this diverse regional context, CESE provides services such as research, technological development, environmental popular education, training, organizational development, communications, and publishing to organizations and rural communities, with the aim of creating a more sustainable society. Of particular importance is the examination and documentation of all projects in which the Centre participates.

CESE currently operates five major programmes, all of which are located in the same area and involve different people and groups such as farmers, women, functionaries, technicians, regional organizations, and local governments. Each project revolves around the conceptual framework of the basic principles, strategies, and educational practices of CESE (CESE 1995).

One project is titled *Participatory Strategy in Research, Training and Organization*. This has supported and consolidated the self-promoted regional organizations of twenty-eight communities, the Regional Organization Against Pollution in Lake Pátzcuaro (ORCA), whose objective is to heal, restore, and protect the natural resources of the area. The second is the *Application and Development of a Model for Sustainable Agriculture in the Pátzcuaro Lake Basin*. This experimental research project is based on the wisdom of *campesinos* (farmers) whose work with the land is representative of the agricultural landscape of the region. The aim of the project is to support, bring back, and innovate ecological agricultural technologies and practices which optimize the cost-benefit relationship, and conserve the natural resource base. The third is the *Human Resources Training for the Development of Regional Sustainability*. The objective of this project, facilitated by community leaders, researchers, practitioners, and technicians, is to design and implement a detailed methodological and thematic environmental popular education programme geared towards the protection and sustainable use of natural resources. The project also strengthens the training processes necessary for citizen action-organizing. The fourth is titled *Citizenship Building*. This project is directed towards the promotion of a regional civil society that is politically and environmentally active in public, and which collaborates with national citizens' movements, programmes for electoral observation, citizens' polls, etc. The fifth is the *Community Health and Organizational Development of Women in the Pátzcuaro Lake Basin*. This project provides women with an opportunity to critically analyze their own experiences, and empower them to bring about change in their own lives. This project is looked at in detail later in this chapter.

CESE Strategies for Building Sustainable Communities

A strategy for sustainable regional development cannot be uniform, but must depend on the environmental and social conditions of each community. Nevertheless, there are some fundamental elements that all strategies of regional and community sustainable development should contemplate.

There was a need for a social-organizational strategy to decentralize governmental power. The success of this approach is dependent on the extent and availability of consolidated capacity, technical support, economic resources, and the structure of grassroots organizations. Without disregarding the important role of the individual in social development within the sustainable development framework, it is regional and community organizations that can express the desire to initiate projects of collective benefit, and help communities to move in a new social and environmental direction.

The economic strategy we began proved to be the most difficult considering the phenomenal inequality within regional economies that profoundly affects the production of rural communities. There is no clear solution to this problem; it lies neither in the return of productive farming groups to a system of self-subsistence, nor in the indiscriminate incorporation of the producers into market production. A balance between these points must be sought that would on the one hand satisfy the basic needs of the family through diverse production, and on the other, generate sufficient income to improve the lifestyle of the producers.

Our ecological strategy was based on the premise that a new ecological rationality can only be constructed from finding new forms of resource use that do not risk the ecological potential of the ecosystem. In the face of such a challenge, the present-day society has two broad options:

- Selective and intensive use of resources according to market demands. This pattern of usage is characteristic of capitalism and is associated with the idea of unlimited growth in the supply of consumer goods. As it has been demonstrated, such patterns are linked to the industrial model and place society in danger of exceeding the limits of growth.
- The multiple and complementary use of resources that does not subscribe to the demands of the market, but to the satisfaction of basic human needs and equal distribution.

Other proposals, such as decreasing production, improving distribution, cost accountability of ecological damage, cleaner production processes, and closed systems of complete recycling can also be considered.

We based our technological strategy on the common assertion in Latin America that there is a lack of comprehensive knowledge in basic science. This is a necessary element for the development of technology that increases the efficiency of natural resources management. Until now, a technocratic view has prevailed—one that claims that experts and sophisticated technology can solve environmental problems. However, efforts have been made to look for alternative technologies that are appropriate and accessible to communities. Advances in this field have yet to be made. It would be absurd to reject all forms of sophisticated technology. Rather, it is beneficial to adopt a cautious and selective approach towards technology that considers and responds to the most urgent needs of the majority of the population. In this regard, even when farming regions are unable to rely on the possibility of an advanced technological contribution, it is not wise to reject technology solely based on one's principles against it. On the other hand, it is necessary to promote the technological knowledge of indigenous people and farmers, no matter how rudimentary it may be, since this knowledge is very valuable to proper management of natural resources in the lake basin area.

We implemented a cultural strategy because it is crucial that different ethnic groups in Latin American preserve their territories and the cosmovisions that nourish their cultures. If this is not done, any effort to preserve customs, traditions, and other cultural features would be demagogical. Indigenous cultural organizations should be targeted to play a leading role in ensuring the continuity of ethnic culture. External organizations can only play a supporting role.

A research strategy is an extremely important component within the project of sustainable development—research to enhance the understanding of regional and community ecosystems as well as their dynamic socio-economic structure. In order to construct environmental knowledge, different types of inter-related research have been defined. These are elaborated through the process of socio-environmental studies and action: specialized and multidisciplinary study by experts in different areas of community and regional sustainable development (biology, limnology, geography, economy, etc.) following theoretical-academic research models, with the added emphasis that the results of the research are didactically socialized in groups and organizations for a contrasting analysis. When accompanied by a more systematic factual analysis and interpretation of socio-environmental phenomena, these studies can be strategically important.

This multi-disciplinary focus perceives the inseparable relationship between the social and natural worlds and establishes that problems in a region can only be solved through (i) the generation and use of high quality scientific information that is reliable and articulate as to natural and social phenomena that operate in the study region; and (ii) wide and determined participation of local inhabitants and base organizations (Toledo et al. 1993, 5). Participatory research carried out by communities, social groups, and organizations has generated:

- knowledge of the present situation by advisers and community workers relating to the social-environmental problem and its integral alternative solutions, combining scientific and grassroots knowledge;
- more dynamic group and organizational processes in order to make changes in present conditions of natural resource use;
- training of directors, leaders, and members of social groups in processes of participatory research;
- environmental agenda in governmental institutions and organizations so that propositions from base associations can be integrated into regional programmes;
- evaluation of governmental programmes and of the organizations themselves; and
- experimentation, demonstration, and extension of eco-technologies.

Our educational strategy, albeit taking into account the need to re-frame formal education, is focussed primarily on linking informal educational processes with sustainable development action. In the context of this option, one is not looking so much for a memorization of content, but critical spaces of open and active dialogue that help broaden factual, physical, biological, social, and economic knowledge, and develop collective understandings of planning and decision-making processes. Training is carried out in the development process of research and planning and their execution, as well as in communication, evaluation, and socialization. This kind of training is organizational, action-oriented, and transformative. First, there must be a certain continuity in subject matter in the curriculum of informal education, because of the objectives and tasks that are proposed by base groups. Second, different functions must be carried out by members of the groups in their areas of organization. And third, theoretical and practical knowledge is necessary for the development of alternative technologies in different areas of socio-environmental action.

Pedagogical Principles and Methods
of Environmental Popular Education

A number of pedagogical and methodological principles which are directly related to environmental popular education are being applied in training programmes for sustainable management of natural resources.

Environmental popular education borrows from the guiding principles of popular education not only its political characteristics, but also its commitment to research, democracy, and participation as an aspect of education. In the Regional Seminar on Community Training for the Management of Sustainable Natural Resources Bulletin (1995), convened by the Network for Popular Education and Ecology (REPEC) of the Latin American Council for Adult Education (CEAAL) and the United Nations Network of Environmental Training (PNUMA), the following heuristic and pedagogical principles were identified:

- A replacement of the traditional teacher-student (educator-community) pattern with a subject-to-subject relationship.
- The development of consciousness-raising in adult education does not occur in a lineal progression, but fluctuates with periods of advances and setbacks.
- Agendas and programmes should be formulated based on situations/problems and not on subject matters. As such, actual demonstrations of examples and concrete actions are preferable, as educational tools, to verbal instruction and textbook reading.
- Using learning and teaching practices to gradually move from the concrete to the abstract, known to unknown, simple to complex, observation to reflection, particular to general, and practice to theory can be renewing.
- Incorporating indigenous methods of motivation building to enhance learning and considering some principles of learning such as the use of new materials, repetition, analogy, continuity, similarities and contrast, and problems with retention and forgetfulness, etc., can be useful.
- From the perspective of environmental popular education, the social-environmental surroundings should be considered as a relevant factor and a knowledge tool. It is necessary, therefore, to resolve conflicts with schedules and school space in order that a creative educational environment can be promoted.

Environmental popular education proposes the organization of educational activities in places that are more conducive to learning than the conventional classroom. Open spaces are ideal for the development, experimentation, and practice of values in relation to real-life situations and problems. Support should be provided where educational processes are institutionalized so that the scope of teaching and learning is not hampered by limitations commonly imposed by institutions. In this way, the false dichotomy between theoretical learning at school and field practice can be overcome. Education within the field of community production can provide simultaneous action and development in formative processes that stimulate new growth and ideas for communal production.

Environmental popular education should emphasize the development of new skills that incorporate technical and productive dimensions to allow communities to move towards environmental sustainability. In this regard, popular environmental education can engender a dynamic transformation that is both social and ecological.

Environmental popular education establishes a respect for distinct forms of skills acquisition related to the cultural characteristics of each region. If each cultural formation has its own ways of making sense of reality, this should be respected and educational processes should be organized around this consideration. In this sense, new disciplines such as ethno-biology have helped considerably in unveiling the profound environmental epistemology of the indigenous cultures on this continent. It is precisely through this field of knowledge that subjects, in particular those in rural situations, can aspire to cognoscitive participation, a form of participation that allows individuals and groups who benefit from educational processes the opportunity to contribute to the theoretical formulation of projects of social transformation. In the context of the case study, the theoretical formulation of this project involves the conceptual and practical construction of sustainable development.

Environmental popular education has the responsibility to support the development of a new environmental understanding. This implies the integration of a fragmented body of knowledge and skills that is necessary to envision an interdisciplinary and cross-sectoral reality. The idea is not to produce experts who are well versed in the ecological crisis, but to develop a global understanding of reality that allows for the crucial identification of social-environmental problems and the initiation of a solution (Reyes 1995).

Women's Community Health and
Organizational Development Project

Indigenous and non-indigenous women in rural communities near Pátzcuaro (and in the entire country) occupy the lowest rung in the social hierarchy.

The number of acts of violence perpetrated against women in the Pátzcuaro region is simply horrifying. In interviews with 500 women it was discovered that only 12 had not been raped. Most were raped by relatives (brothers-in-law) or by ritual kin (godfathers or fathers-in-law). In literally all the cases, no criminal charges were filed.

Male migration to urban areas has compounded women's workload by increasing their need to find paid work while continuing to care for the large numbers of children they have as a result of religious prejudices or lack of information about contraceptives. Women's cosmo-visions are partly governed by the theological principles of the Catholic Church and indigenous materialist interpretations, which are evident in their ideas and explanations about diseases and health. Ideologically speaking, these women have come to assume that inferiority is their destiny and any modification to this belief system is unimaginable.

Even though environmental deterioration is a condition that affects everyone, it should be noted that women suffer the most from it because of their subordinate position. The consequences of the resultant increase in their workload are reflected in daily practices such as the collection of firewood. Scarcity of firewood means that longer distances must be covered for collection. Women must also travel farther in order to obtain basic food products for their families since fruits and vegetables are no longer produced in their own gardens. According to the same women, the difficulty they have controlling garden pests has made food cheaper to obtain from the market. This means that women must now travel to the nearest urban centres to purchase food items (for examples, see Clover 1996; Rodda 1991).

Additionally, aggressive corporate advertising of low-cost, low-nutrient processed food, and the increased consumption of these products in even the remotest communities are resulting in severe problems of malnutrition. This malnutrition and the consumption of contaminated water are two of the key factors behind rapidly deteriorating health conditions in the communities. In their daily lives, women are exposed to these situations more directly and, therefore, any action to solve these problems must include them.

In the past, a different relationship between humans and nature existed; a relationship which was more balanced and reciprocal. While this is still the case in indigenous communities, successive generations are gradually losing their traditions, customs, and beliefs as influences of modernism and post-modernism encroach ever more rapidly (Cira 1996).

For all the reasons listed, the primary participants in the health and organizational development project are women from rural lakefront communities in the Lake Pátzcuaro area. The animators of this project are women, personnel from the CESE who come from the same rural communities. At present, nine communities are involved in intensive action-oriented educational projects.

Training of the animators combines action strategies, educational methods, and social organization with various themes or issues in the community. The training process includes on-the-ground work supplemented by short courses, workshops, and seminars, both within and outside of the project.

Objectives and Principles

Project Objectives

The project came into being as part of CESE's regional and local programme with the following objectives:

- Promote regional sustainable development through working with women;
- Stimulate community and regional organization with the participation of Purépecha women; and
- Provide the space and information for women of the region to analyze their reality critically, and to understand their rights.

Consequently, the project is based on a variety of principles: The first is the strategic relevance of grassroots work to the family, which complements regional organizing projects supported by CESE; second, the ethical and political commitment toward disadvantaged sectors of society. It is recognized that rural women are more oppressed than their male counterparts; last, the impact of the feminist movement in demonstrating the relevance of women's role in social and environmental transformation. By integrating the gender imperative, this environmental paradigm takes into account the relationship between men and women as the basis for the analyzing social relationships and their interconnectedness to nature (Toledo et al. 1993).

The Project Strategy

The strategy of this project has evolved during its seven years of existence. The first phase of the project was the Institutional Set-Up. At this stage, the project team elaborated the basic features which constituted the project. Funding was also obtained from the Dutch Humanist Institute for Cooperation with Developing Countries (HIVOS), in order to begin the project. The project took roots with the participation of two professional women. The second phase was the initial community group contact. Previous work in the region allowed the project team to quickly establish contact with groups of craftswomen in two communities. These women participated in activities for the sale of their products. However, since the problem of sales was somewhat resolved, CESE re-focussed its support to more pressing issues. A third phase was based on women's self-analysis. In response to the women's concern, a self-analysis was carried out which showed that health issues were a big priority due to the high cost and negative side effects of allopathic medicine in the treatment of microbes and related diseases. The fourth phase was the health project. From the results of the self-diagnostic, the women initiated health-oriented activities from a curative and preventive standpoint. As the first objective, each group worked to recover their indigenous herbal medicine which was crucial to the revitalization of traditional knowledge about plants and their medicinal properties. The women conducted their research through social interaction, visits to the countryside, and interviews with older women. The learning process did not end with the recovery of traditional knowledge, but with the incorporation of new techniques, such as the production of plant extracts and micro-doses. The research also included a biological dimension with the participation of botanists, who collaborated in the scientific classification of the plants in order to extend knowledge of their use outside of the region. In this way, an exchange of knowledge between the community and the researchers is maintained. The fifth phase was the Basic Supply Project. Since nutrition is a health-related issue, and because of the high prices of basic goods for the home, work has been carried out with women's groups to provide training in the planning of basic balanced diets based primarily in the recovery and more extended usage of regional products. At the same time, the women have organized themselves to obtain basic supplies. Different methods of organizing were selected according to the characteristics and needs of each group. The most efficient method was one in which the groups created a list of what was needed and these items were distributed to each person fortnightly. The collective store is another model even though the groups have generally preferred to maintain a system that is less risky and that does not change their daily life patterns too drastically (since a store implies a simultaneous commitment to new financial, organizational, and administrative responsibilities). A community store run by fishers is now responsible for supplying the needs of their community and for distributing basic food products to the women's groups in the community. This phase was followed by an ecological strategies phase. The women's groups have been powerful regional organizers of the community, principally through their involvement in assemblies and activities of the Regional Organization Against Pollution in Lake Pátzcuaro (ORCA). In regional working plans, specific tasks have been identified to prevent deforestation and to enable environmental healing. Women's groups have assumed an important role in adapting technologies, particularly those related to domestic space. For example, there has been direct participation in the construction of a large percentage of the 500 improved mud stoves that have been built in the region. These stoves consume up to 50 percent less firewood than the unimproved stoves. In fact, the transmission of this ecological technique has been horizontal, that is, from woman to woman. Another phase was to foster inter-group relationships. One practice that has been maintained throughout this project is a regular dialogue and evaluation of the progress made by group projects. These inter-group sessions are held once every three months and they will lead to the creation of a permanent regional women's forum, and the formation of a regional women's organization. Finally, an explicit analysis of the various levels of relationship between men and women (husband-wife, father-daughter) requires a lengthy process of sensitization. The strategy that has been adopted by the project has been effective in that women can express their own views about marital problems, a subject which they were formerly extremely uncomfortable about discussing in public. The trust gained by the project team evident from the efficient training in health issues, advances made in the basic supplies project, and the relationship between the patients and the project head (a female doctor) have made discussions of previously 'offensive' issues possible at the group meetings.

Reproductive health issues are dealt with in reflection workshops where daughters-in-law are separated from their mothers-in-law. Commonly within these communities, the mothers take charge when their sons are absent, and in many cases they treat their daughters-in-law severely, clearly demonstrating that the status of a mother-in-law lies in the power she has gained through the passage of time. Domination of women by women is therefore the first issue to be discussed with both generations of women (mothers-in-law and daughters-in-law). As women begin to understand their own bodies, they are able to voice their analysis of women's public and private relationships. In this way, a biological and cultural vision of the relationship between men and women, and the relationship between humans and nature, can be integrated.

Conclusion

Due to escalating environmental problems, CESE has implemented a variety of environmental popular education activities throughout the region. However, we have learned well that any educational process that claims to work towards a more healthy and sustainable community must by necessity include the transformation of present relations between men and women. It must also work to empower women, giving them voice, and acknowledging their wisdom and needs.

Any attempts to build a sustainable society without the full participation of women is doomed to fail because it lacks the ethical, political, and cultural bases that can transform the dominant relationship between society and nature. Women's participation is necessary in the reclamation of their rights, and to open up the possibility of global decision making in politics and practices for the creation of sustainable communities.

References

Barrera-Bassols, N. 1993. "Ecogeografia." In V. Toledo et al. (Eds.), *Plan Pátzcuaro 2000*: *Investigacion Multidisciplinaria para el Desarrollo Sostenido*. Mexico: Fundacion Friedrich Ebert.

Bosco, Pinot Joao. 1977. *Extension o Educacion. Una Disyuntiva Critica*. Santo Domingo: Instituto Interamericano de Ciencias Agricolas.

Centro de Estudios Sociales y Ecologicos (CESE). 1995. *Documento Sintesis sobre Programas Institucionales*, CESE. (Mimeo). (Parts of this article were published previously under the title "La Educacion Popular y la Formacion del Saber Ambiental" by Joaquin Esteva and Reyes Javier, in "Materiales para la Formacion de Educadores Desde la Educacion Popular, No. 2." Santiago de Chile: CEAAL.

Centro Regional de Educacion de Adultos y Alfabetizacion Funcional (CREFAL). 1980. *Informe Final del Taller sobre la Metodologia de la Investigacion Social en los Programas de Capacitación Rural.* Pátzcuaro, Michoacan: CREFAL.

Chacón, A. 1994. *Pátzcuaro. Un Lago Amenazado.* Mexico: Universidad Michoacan de San Nicols Hidalgo.

Cira, Y. 1996. *La Mujer Como Agente de Cambio en los Problemas Ambientales.* (Mimeo). Pátzcuaro, Michoacan: CESE.

Clover, D. 1996. *In Their Spare Time: Women and Environmental Action.* Conference Proceedings of the 1996 National Women's Studies Association. Oklahoma: NWSA.

De Schutter, A. 1996. *La Investigacion Participativa. Una Opcion Metodologica para la Educacion De Adultos.* Pátzcuaro. Michoacan: CREFAL.

Esteva, J. 1995. *The Restoration of a Lake Basin in Mexico and the Regional Social Participatory Process. The Case of ORCA.* Center for Social and Ecological Studies, A.C. Pátzcuaro, Michoacan, Mexico: World Resources Institute.

Hall, B. and Gillette, A. 1977. *Participatory Research.* Toronto: International Council for Adult Education.

Leff, E. 1992. "Hacia la Construccion de una Nueva Racionalidad Ambiental." *Pachamama,* No. 5, Red de Educacion Popular y Ecologia (REPEC), Santiago de Chile.

Red de Educacion Popular y Ecologia (REPEC). 1996. *Programa de Naciones Unidas para el MedioAmbiente (PNUMA).* Seminario Regional sobre Capacitacion de las Comunidades para el Manejo Sustentable de los Recursos Naturales. Informe, REPEC, Pátzcuaro, Michoacan.

Reyes, J. 1994. "America Latina. Ambientalismo y Educacion." *Educacion Popular Ambiental en America Latina,* REPEC, Morelia, Michoacan.

Reyes, J. 1995. *Conclusiones de los Talleres sobre Formacion de Pomotores Ambientales del Colectivo Mexicano de Educacion Popular Ambiental,* Mexico: CESE/REPEC.

Rodda, A. 1991. *Women and the Environment.* London: Zed Books.

Toledo, V., Barrera, N., and Avila, P. 1993. *Plan Pátzcuaro 2000.* Mexico: Fundacion Friedrich Ebert.

Part III

Community Practice, Reflection, and Research

Chapter 8

Facilitating Community Environmental Adult Education in the Philippines: Transformation in Action

José Roberto Q. Guevara
Victoria University, Melbourne

We showed them photographs of a forest.
spears of sunlight piercing the blanket of leaves
trees stretching their limbs towards the heavens
We listened to nature sounds from a cassette tape.
the gushing river cutting through the forest
the dancing cicadas, singing their nightly tunes
We read the latest statistics.
less than a million hectares of old growth forest
more than eighteen wildlife species in the endangered list
They saw,
they listened,
they read,
but had they experienced?

We[1] were surprised by the reaction of an urban-poor leader and educator whom we consulted about a proposal to develop an environmental adult education module for the urban poor. While he said that people might be willing to attend a training workshop, they would have to be convinced of the long term contribution of environmental issues to community concerns around the demolition of homes, health, and poverty. As she said, "Environmentalists work for a clean and green environment. Since we live around garbage dumps and some of us survive by scavenging, environmentalists would look at us as part of the problem."

So, at the start of the training workshop, we thought we should go to the Makiling Rainforest Park, a two-hour drive from Manila, for some awareness and sensitivity activities that could give the participants time and opportunity to interact with nature. The sharing that followed was inspirational. This was the first time many of the participants had embraced a tree, breathed clean air, listened to the symphony of birds and crickets, and felt and tasted cool, clear water. The park was an escape from the smog, the noise, the stench, and the pace of the urban environment.

We took this situation as a challenge as we realized that few urban poor have ever experienced a clean and healthy environment. Living all their lives in the depressed areas of Manila is usually a life beset by air, water and solid waste pollution.

I wrote the lines at the beginning of this chapter in my reflective journal after a workshop. A few weeks later, one of the adult educators from the community visited our office to collect materials and a video on the current state of the Philippine environment for an upcoming workshop he was conducting. Looking back, I think the day in the rainforest was a moment of learning, a moment of transformation.

This chapter describes the different transformative learning moments experienced by the Center for Environmental Concerns—Philippines (CEC) in the process of developing an environmental adult education programme with local communities. CEC is a non-governmental organization committed to environmental protection and rehabilitation. Established in 1989, its role has evolved from providing basic environmental adult education to facilitating pro-active environmentalism based upon community-based environmental assessment and monitoring, education, training, advocacy, and organizing.

The chapter weaves four transformation stories. It begins with the transformation, for the worse, of the Philippine environment. This is followed by a reflection on the transformation of CEC's awareness to environmental action workshops called RENEW (Restoration Ecology Workshops), and its contribution to shaping a grassroots environmental adult education programme in the Philippines. The third is my own story of transformation as a community environmental educator. The chapter ends with the ongoing story of transformation which we environmental adult educators experience as we become active partners of local communities, in our struggles to address the root causes of environmental problems.

The Philippine Movement: A Story Waiting to be Transformed

The Philippines are rich in stories of how we have been blessed by nature with 7,100 islands, fertile soil, and clean water which support a rich diversity of plant and animal life. Sadly, we are experiencing a very different story today. In less than thirty years, the more than 16 million hectares of forests recorded in 1968 has been reduced to a mere 5.6 million hectares. From 1989 to 1995 the average rate of deforestation was 130,000 hectares per year. A 1993 study reported that 5.2 million hectares of the country are severely eroded (EMB-DENR 1996). Fifty of the 421 rivers in the country are considered biologically dead. Only 120,000 hectares of vegetated mangroves remain from the original 500,000 hectares in 1920. Only 5.3 percent of coral reefs are considered to be in excellent condition while eleven out of fifty major fishing areas suffer from over-fishing (CEC 1996). By themselves, these statistics are staggering, but the condition is exacerbated by natural and human-induced calamities. Together, the accumulated impact of these catastrophes on the resource base for people's livelihoods has been alarming.

The drive to transform the Philippines into a Newly Industrialized Country (NIC) further endangers these resources. The rate of conversion of agricultural land to industrial zones, residential subdivisions, or recreational facilities such as golf courses continues to increase. Sustaining this pace of industrialization will require the construction of coal-fired thermal plants and geothermal plants to supplement the country's oil-dependent energy generation plants. The exploration and extraction of mineral resources by foreign-owned mining companies and the accompanying social and environmental problems continue. All these continue to push the lives of already marginalized populations to the brink of desperation.

The effects of these trends include the increased migration of Filipinos from the rural areas to the major cities looking for work, and the subsequent problems of over-crowding, solid waste pollution, flooding, and the spread of communicable diseases.

While there has been a relative increase in the level of awareness and action among the different sectors of Philippine society, the task of challenging the direction and checking the rate of industrial development in terms of sustainability goals remains. Environmental adult education contributes to this task. As environmental adult educators, we can work together to facilitate processes that will help our people rewrite these sad contemporary stories about the environment.

Renewing RENEW: Educational Transformation

RENEW—or the Restoration Ecology Workshop—was CEC's initial attempt at planning workshop experiences for a community-based programme. RENEW can be described as an Awareness to Environmental Action Workshop with the following objectives:

- to heighten environmental awareness by clarifying the participants' realization that they are part of a larger ecosphere currently threatened;
- to promote an understanding of basic ecological concepts;
- to analyze the scope, causes, and impacts of the current environmental situation in their respective sectors and/or localities;
- to develop concrete responses to the environmental problems identified.

RENEW workshops have the following themes or modules:

- Acquaintance, Release, Expectations Check, and Orientation
- Basic Ecological Concepts (ecosystem structure and function; balance of nature)
- State of the Environment (sectoral, local, national, and global)
- Vision Towards Environmental Action
- Evaluation and Celebration

Structurally, RENEW workshops are designed with the following conceptual learning map in mind:

AWARENESS ———> UNDERSTANDING ———> ACTION

Matching the content modules with the above learning map results in the following sequence of modules:

Awareness and Understanding

- Acquaintance, Release, Expectation Check, and Orientation
- State of Local Environment
- Understanding

- Ecosystem Concept
- Ecosystem Structure and Function (Energy Flow and Chemical Cycling)
- Balance of Nature
- State of National and Global Environment

Action

- Vision Towards Environmental Action
- Evaluation
- Celebration with Nature

A participatory approach to environmental adult education is used in RENEW workshops. This approach recognizes that we affect and are affected by our interactions with our environment, and that we should participate in finding solutions to the problems we experience. We all share a responsibility for finding solutions regardless of whether we are scientists, fisher folk, politicians, indigenous peoples, farmers, policy makers, workers, teachers, students, women, or children. This perspective challenges the notion that environmental problems are technical problems requiring technological solutions. In reality, they are much more complex.

RENEW provides an opportunity for people to share their experiences and learn from each other about how the environment functions, and what they can do, individually and collectively. To further enhance the process of sharing and learning, creative educational approaches are utilized with local facilitators trained to ensure the decentralization of CEC's environmental adult education efforts, and the development of a locally relevant programme.

The Restoration Ecology Workshops have lived up to the acronym RENEW because they continue to be renewed prior to, during, and after each workshop. The Nature Awareness and Sensitivity module for the workshop described at the beginning of this chapter is an example of such ongoing changes. We called this 'progressive contextualization': the ongoing process of adjusting learning objectives and training modules, revising frameworks and approaches, and reflecting on practice and experience in response to the particular context of the learners. This process not only renewed RENEW but also renewed us as community environmental adult educators.

Sectoral and Curricular Integration

Most of RENEW's growth comes from working with sectoral groups through federations or networks of women, farmers, fishers, the urban poor, workers, and indigenous peoples. Prior to 1991, RENEW workshops were predominantly multi-sectoral workshops conducted at a regional or provincial level. In 1991, we initiated the Sectoral Environmental Adult Education (SEED) Program to develop sector-specific environmental adult education modules and adult educators. This was seen as our contribution towards the strengthening of the former Philippine Environmental Action Network which facilitated the development of the Peoples' Environmental Agenda.

A RENEW trainers' workshop was conducted with adult educators from various national federations, to identify sector-specific environmental concerns and to develop a basic environmental adult education module that would be implemented jointly with adult educators from CEC. Most of the modules designed during this training workshop could be called RENEW Plus as they are the result of advice from the adult educators in these organizations, to simply add a new topic to address the specific environmental concerns of their sectors. Thus, we introduced workshops with topics such as sustainable agriculture for farmers, gender and environment for women, genuine aquatic reform for fishers, and occupational health and safety for workers. However, a danger with such an approach is that environmental issues could be perceived to be distinct from the current concerns and demands of these sectors.

A breakthrough came when adult educators from the women and agricultural workers sectors suggested that we explore the possibility of incorporating environmental awareness objectives into their existing basic courses, rather than creating a new course. This initiative facilitated the integration of environmental issues within their sectors' issues. As a result, other adult educators have sought to incorporate environmental topics into the following basic courses: "Genuine Trade Unionism," "Gender," "Reproductive Rights," and "Occupational Health and Safety." For example, the concept of ecosystem has been used to facilitate an understanding of the complex interrelationships of environmental problems with the historical and current socio-economic, political, and cultural conditions of our country, and to discuss the connections between the biological and the physical components of the environment. Similarly, the basic module on the national situation, often limited to a discussion of the socio-economic and political situation, was expanded to include a brief discussion of major environmental problems. The integration of environmental issues into sector-based education activities was just the first step. The bigger challenge, which continues as I write this chapter, is to facilitate the integration of environmental concerns into the specific agenda of the different sectoral federations.

Dual Learning

It is important to acknowledge the dual learning process that occurred within the SEED Program; especially in relation to the steep learning curve we experienced working with indigenous peoples and in facilitating the integration of women and gender concerns.

Three critical issues about our practice of environmental adult education arose while working with adult educators and leaders from indigenous peoples and groups. First, the concept of studying the environment was foreign to them. They told us that their lives are so intimately connected with the environment that learning is an ongoing process that happens with, and in, the environment. According to them, we, the urban-based individuals, were the ones who needed to develop a deeper understanding of our relationship with the environment. This feedback itself made us more aware of the different assumptions that we as environmental adult educators bring, especially when we work with groups from different learning and cultural contexts.

Second, the indigenous participants said they enjoyed the creative learning methodologies but were uncomfortable with the approach that seemed to 'make fun' of the environment, which they regard as sacred. They believed that we could gain as much knowledge about the environment sitting around the campfire and sharing stories, as playing games. This comment drew our attention to the need to explore ways of learning that may be more appropriate and sensitive to different cultures, without compromising our commitment to participatory and creative methodologies.

The third issue concerned the basic concept of ecosystem structure that states that the ecosystem is composed of living and non-living things. The indigenous groups found the idea that there are non-living things to be unacceptable. "Water is alive. Rocks are alive. The Sun is alive." This point of view was also shared by the farmers and fishers with whom we worked. They argued that the soil and the sea are alive because they support both plant and animal life. The biology teacher in me wanted to argue that water, rocks, the soil, and the sea could be defined as ecological systems made up of living and non-living things. The Sun, on the other hand, could be seen as the ultimate source of energy that sustains life. Instead of challenging them on these beliefs we learnt to listen and tried to understand their perspective.

We came to realize that understanding the concept of interdependence does not require a classification of living and non-living forms, and we started to explore how this particular view of nature affects the kind of interactions people have with the environment.

The majority of the adult educators we train are women in a wide range of sectors, and more than half of the CEC staff members are women. However, gender is a concern that we continue to struggle with. To cite one example: In 1993, after months of planning a comprehensive programme for the first Advanced Trainers' Training, some of the participants, on the second day of the programme, said that we needed to have a discussion on women and gender before the training could proceed. We agreed to the request and asked a couple of the women to facilitate the session.

It did not take long for us to realize that the reaction was a result of the use of offensive sexist language by some of the participants. In true adult learning fashion, the women facilitators used this situation to launch a discussion of women's oppression followed by an attempt to link this to their role as environmental adult educators. There was not enough time to deepen the discussion, but there was an agreement to stop the use of sexist language. It was a brief session but it helped us become aware that we were gender-blind when we designed the training.

During this same year CEC joined a task force of women's organizations that facilitated the development of a framework on Women, Environment, and Sustainable Development. This was followed by a short-lived project called "Women Community Action Planning for Rehabilitation" that developed preliminary modules on women and environment based on the framework. In 1998, a project on women and mining was implemented, training local women to become researchers. This project documented the impact of mining on their lives. However, the center experienced difficulty in employing a gender perspective in facilitating an analysis of the documented experiences.

A review in 1998 highlighted some of the causes of the slow growth of women and gender concerns within the programmes of the center. While we were open to the issue, we realized that the impetus was coming from outside the organization, specifically from some funding bodies and women's groups. These important concerns were not rooted within the organization as gender was not an issue identified as a priority by local partners, nor was it a strongly felt need by the majority of the center's staff. This is an example of a gender-blind perspective that tends to perpetuate itself.

These stories of our learning from indigenous and women's groups are just two examples that have made us aware of the assumptions that had found their way into our work. Like RENEW, we continue to strive to challenge these assumptions, transforming awareness into action in our own lives and activities. As former CEC executive director, Noel Duhaylungsod wrote the following in the preface to *Renewing RENEW: A Restoration Ecology Workshop Manual* (Guevara 1995).

> The basic concepts of environmental science could be learned by anyone because they can be drawn from real life experiences. The environmental milieu is inclusive of the biophysical, social, cultural, economic, and political environments. As such, environmental adult education modules must correspond to the local ecology and the situational specificity. It is crucial that the education module and the curriculum must be progressively contextualized. And that method or process is as important as content and context.

In 1990, we dreamed of developing a curriculum for community environmental adult education: RENEW was our first attempt. Renewing RENEW continues.

While RENEW is still considered our core programme, in terms of content and process, it has provided the seed for the development of many other environmental adult education programmes. The Awareness to Environmental Action Series now consists of the following training workshops:

1 Environmental Orientation—a half-day or full-day workshop that covers the topics of RENEW on a general level.
2 RENEW—a three-day live-in workshop. Trainers Training for Participatory and Creative Approaches to Environmental Adult Education—often conducted after a RENEW workshops and taking two to three extra days.
3 Advanced Trainers' Training—a weeklong workshop that primarily is a venue for sharing environmental adult education experiences and upgrading relevant skills such as environmental monitoring and participatory research methods.

4   Community-Based Environmental Monitoring (CBEM)—a series of training workshops that provide skills in bio-physical, socio-economic, and health monitoring of a specific industry (e.g., a gold mine, copper smelter, etc.) or an infrastructure project (e.g., road construction or land conversion).

5   Community-Based Rehabilitation Technology (CORETECH)—a series of workshops that develop knowledge and skills for specific rehabilitation projects (e.g., agroforestry farming to prevent erosion, or duck-raising to counter the problem of the Golden Apple Snail infestation in rice fields).

Issue-specific and skills-focussed workshops are also designed in response to requests from local communities or organizations (e.g., potential impacts of mining and conducting an environmental investigation mission or risk assessment).

Educational support materials have also been designed and tested, and a regular newsletter, called *FEEDBACK,* audio-visual materials, and a series of worksheets called *Understanding Our Environment* in the three major languages spoken in the Philippines (English, Filipino, and Bisaya) have been published. *Renewing RENEW: A Restoration Ecology Workshop Manual* documents CEC's experiences in grassroots environmental adult education and was published in 1995. It describes and analyzes the development of the RENEW programme, including the various learning activities that were developed and utilized during workshops.

RENEW has also assisted in the formation of local and regional environmental organizations and networks that worked with the former Philippine Environmental Action Network. In 1996 CEC facilitated the organization of the People's Faculty of Grassroots Environmental Adult Education and Studies, a conferential body of community adult educators that works in partnership with CEC in developing and implementing localized environmental adult education modules.

Reflecting on Practice

Like RENEW, all our activities follow the principles of participatory and creative education in both content and methods. Years of experience have deepened our resolve to apply these principles to environmental adult education, and to incorporate into it elements that would make it experiential, contextualized, evocative, scientific, critical, and action-oriented. What follows is a brief discussion of these elements we have identified from our practice, that resonate with the work and writings of other adult educators.

The principle of participation recognizes that adult participants have their own knowledge and skills gained from life experiences which they can contribute to the learning process (PRIA 1995). On the other hand, the educator also contributes his/her own knowledge and experiences in facilitating this participatory learning process (Wilson and Burket 1989). Therefore, everyone assumes joint responsibility in a dialogical learning process. The difficulty we often experience at the start of most workshops is how to encourage participants to acknowledge their current knowledge and skills, particularly those participants with minimal or no formal schooling. Beginning the learning process from the participants' life experiences has helped us address this difficulty and allowed us to challenge assumptions about knowledge and its production. These assumptions often create a subtle power imbalance between the learner and the educator, most commonly manifested in a one-way flow of knowledge from the 'teacher' to the 'student.' As environmental adult educators, we need to be aware of these invisible power structures and related assumptions that exist in the learning environment. These are very important to acknowledge and challenge when the 'educator' and the 'learners' come from different contexts. In particular, when we enter a community-based learning setting, we need to be transparent about our objectives, expectations, and the experience and knowledge we bring, and situate these in the context of the participants, their community, and their issues, to facilitate a more participatory learning process (Arnold 1991).

Such a process encourages participation but also underscores the value of experiential learning in adult education. Dewey, as early as 1938 (cited in Wilson and Burket 1989), recognized that traditional education tends to impose the subject matter on learners, without the corresponding experience to understand it. He recommended that education be embedded in the real life experiences of the learner. This is equally true for environmental adult education, which acknowledges that we are all part of, and interact with, the environment. This makes our most mundane everyday experiences a rich source for learning about the environment.

Caine and Caine (cited in Knapp 1992, 4) supported this view when they stated that "all learning is experiential." They also suggested that adult educators must select experiences that are appropriate and meaningful to the learner. Therefore, environmental adult education should be contextualized, that is, based on an understanding by the educator of the conditions of the learner. These conditions often cut across the various levels that the learners are immersed in or wish to become aware of—from the personal, local, national, regional, and at times, global context. Our experience shows that such contextualization is a dynamic process, hence, a progressive contextualization (Guevara 1995).

An additional challenge lies in the development and use of methods to assist in drawing out these relevant life experiences from the participants. This evocative character of environmental adult education has to overcome what Freire (1970) calls the "culture of silence," which often can be observed in adults trained by an educational system that promotes a one-way flow of knowledge, and further disempowered by a socio-economic and political system that oppresses them. Using an evocative method for learning is therefore consistent with the participatory and experiential nature of environmental adult education.

The use of creative approaches, often equated with songs, games, role-plays, and simulations, facilitates this evocative process by actively engaging the participants in the learning process. In certain instances, these methods facilitate drawing the participants slowly into the learning process by overcoming their inhibitions. However, it is important to ensure that the methods used are culturally sensitive. For example, an activity that may require men and women participants to hold hands may be inappropriate in some cultural contexts.

Breaking out into an action song about the adverse impacts of logging may help keep the energy high and add variety to what might be a very technical discussion. However, care must be practised in that the creative activity does not overshadow the intended content outcomes. In addition to creativity in methodology, the extent of environmental degradation we face calls us to think creatively in terms of potential alternatives and strategies for dealing with environmental problems.

Creativity complements the need for environmental adult education to be scientific in both process and content. By scientific, we mean that the learning, like the scientific method, begins with an observation, which may lead to asking some questions (problem identification), looking for preliminary answers (hypothesis and experimentation), followed by developing a deeper awareness of the initial observation (analysis and findings). This is similar to reflexive observation, a major tool in active learning, identified as a part of the learning cycle that Freire and others have described.

In terms of a scientific content, relevant ecological and social science concepts are studied to assist in analyzing the observation and in identifying possible ways to address the problem. This, however, does not imply that scientific knowledge is the only acceptable tool for analysis, since environmental adult education is open and encourages the cross-fertilization of other forms of knowledge, such as indigenous and local knowledge.

While we value and often begin with local knowledge and experiences, this knowledge, like other forms of knowledge, needs to be the focus of critical reflection and not accepted without question. Mezirow (1981) emphasised this but also made sure that critical reflection helps both adult educators and learners to become conscious of their own knowledge, and to be critical of it. Our experiences with indigenous peoples and women are concrete examples of the need for critical reflection, not just at the level of awareness but also at the level of concrete action.

Environmental adult education is action-oriented in process and outcome. This combination of experience, reflection, and action can be traced to the CEC's ongoing practice of participatory action research (McTaggart 1989). This gives environmental adult education the transformative potential for facilitating environmental change at various levels.

Transformation Story

I decided to include my story to provide an example of the personal learning (and unlearning) processes that were occurring while the curriculum was evolving. Two threads weave through my story as a community environmental educator. The first involves the tension between teaching and learning, particularly during the early design stage of RENEW. Biology and ecology were my majors at university. I then taught them for five years at both secondary and university levels. This experience preconditioned me to think that community-based environmental adult education meant teaching the concepts of ecology that could then be applied by communities in their daily lives. Looking back, it is understandable that the list of concepts I selected read like the course outline of Biology 150: Principles of Ecology I, a subject I taught in the university. It did not take long for my colleagues to point out that while the outline was exhaustive, it was not appropriate in terms of its scope, level of detail, or usefulness.

Streamlining the outline involved identifying core concepts for developing an awareness of, and actions for, the environment, based on an understanding of the expected participants and their local contexts. We decided that the concept of interrelationships between components of the environment would be the core concept. This concept had to be made concrete in terms of the functional relationships in the ecosystem, hence the inclusion of topics on bio-geochemical cycling and energy flow. Finally, all these had to be linked to the concept of balance and sustainability within the context of ecosystems and the broader socio-political structures.

This is a far as we could go without completely pre-empting the influence of the participants on course content. However, as soon as we tested the workshop with our first local trainers, the nature of the interaction shifted from teaching ecology to learning and sharing life experiences about the environment with local communities.

The initial curriculum development process was described by an evaluator, during the 1993 RENEW evaluation, as prescriptive because the designers determined what was to be learned. We called the eventual approach 'prescripatory,' to indicate the combination of prescriptive and participatory approaches that balanced the roles of learning and teaching in the development of a community-based curriculum.

A second tension was between my scientific training and the creative approach to learning. My experience as a community theatre artist and educator convinced me that there was a more effective approach to environmental adult education than the conventional lecture, small group discussions, and the occasional audio-visual presentation. The use of a creative and participatory approach was often restricted to energizing weary participants after long sessions. RENEW tried to integrate this energizing function with an attempt to help focus the participants' attention on the topics being discussed. This facilitated the development and adaptation of a good number of games and action songs that dealt with environmental issues.

In our enthusiasm, we eventually tried to squeeze too many of these activities into a three-day workshop. Some adult educators call this the 'feel-good workshop' syndrome, where the enjoyable creative methodology ends up overshadowing the workshop content, instead of helping to focus the participants on the learning objective. After the initial excitement, we became more selective as to when and where to use these activities.

It was also important for us to distinguish between developing the ability to think creatively and the use of creative activities. While the latter helped make learning enjoyable, the former facilitated the development of a positive outlook towards the complex environmental problems they faced.

These tensions did not require me to make a choice. After all, it is not a matter of academic knowledge or local knowledge, nor is it about being scientific or being creative. These tensions need to be balanced as part of the process of developing community environmental adult education that is holistic in scope, context-based, and action-oriented. As an educator, my own context has influenced and been influenced by this process. I continue to weave these threads and more. I now prefer to view these as creative tensions, an essential part of my transformation and growth as an environmental educator.

Transforming Ourselves as Environmental Adult Educators

We are often asked, as adult educators, what impact has our work had on our communities? Have we transformed their lives? Have we transformed their relationship with nature? Have we found solutions to address their environmental problems? From my perspective, two factors make these questions difficult to answer. First, community education outcomes are difficult to gauge. While they are defined by specific objectives, the level of learning is determined by how the knowledge gained is adapted and applied to different problems and situations, and not by quantifying what is retained through mere memory. Therefore, all community environmental adult education work needs long-term monitoring, as well as ongoing support for learning and action. Second, such questions seem to place the burden of transformation on the educator. But education is only one component within the wide spectrum of environment work; in fact, only one component in the entire process of community development and empowerment. Adult educators can facilitate learning; it is the local people who ultimately hold the capacity to transform themselves and their situation. This recognition must guide the basis of transformative learning—to empower people to believe in their capacity to change themselves, their community, and their environment.

How then can we, as environmental adult educators, facilitate this lifelong process of learning and transformation? When CEC decided to be a national technical service center on environmental concerns, we accepted and worked within the limitations of not being based at the local community level. We saw our involvement as primarily to train local community adult educators, and assist them in the development of local environmental adult education modules and materials.

To date, we have assisted in the formation of local and regional environmental groups, facilitated the integration of environmental concerns in sectoral education curricula, trained local environmental adult educators, and established the People's Faculty of Grassroots Environmental Education and Studies.

One significant realization from this experience is that we are not just catalysts for change. Catalysts, by definition, hasten chemical reactions but are not transformed by these processes. Our experience has shown that as facilitators of transformative learning, we ourselves need to be open to being transformed. What a humbling, yet empowering, realization.

Note

1   I use the pronoun 'we' here in order to acknowledge the contributions of Melissa Morales, Mercedes Llarinas, Cyril Millado, Queenstien Banzon, and Delailah Padilla, who were all involved at various times with the Education and Training Department of the CEC–Philippines. But most of all, I wish to acknowledge the commitment and contribution of the sector-based and community-based educators, our partners, without whom all these transformative learning experiences would not have been possible.

References

Arnold, R., Burke, B., James, C., Martin, D., and Thomas, B. 1991. *Educating for a Change.* Toronto: Between the Lines.

CEC. 1996. *A Primer on Philippine Environmental Data.* Quezon City, Philippines: Center for Environmental Concerns.

EMB-DENR. 1996. *The Philippine Environmental Quality Report,* 1990–1995. Quezon City, Philippines: Environmental Management Bureau—Department of Environment and Natural Resources.

Freire, P. 1970. *Pedagogy of the Oppressed.* New York: Herder and Herder.

Guevara, J.R. 1995. *Renewing RENEW: A Restoration Ecology Workshop Manual.* Quezon City, Philippines: Center for Environmental Concerns.

Knapp, C.E. 1992. "Lasting Lessons: A Teacher's Guide to Reflecting on Experience." ERIC Clearinghouse on Rural Education and Small Schools, Charleston, WV.

McTaggart, R. 1989. "Principles of Participatory Action Research." Paper presented to the Third World Encounter on Participatory Research, Managua, Nicaragua.

Mezirow, J. 1981. "A Critical Theory of Adult Learning and Education." *Adult Education 32,* 1, 3–24.

PRIA. 1995. *Training of Trainers: A Manual for Participatory Training Methodology in Development.* New Delhi: Society for Participatory Research in Asia.

Wilson, A. and Burket, L. 1989. "What Makes Learning Meaningful?" Paper presented to the Annual Meeting of the American Association for Adult and Continuing Education, Atlantic City, NJ, October.

Chapter 9

Critical Environmental Adult Education: Taking on the Environmental Challenge

Karen Malone
RMIT University, Melbourne, Australia

Introduction

If we have learned anything over the past thirty years, it is the fragility of our Earth. The growth of industrialization and the demands it stimulates are, in the interests of a small minority, destroying the environment and undermining the livelihoods of millions of people. Alongside this environmental destruction is the increasing impoverishment of the Earth's inhabitants. Socially, culturally, and environmentally our planet is becoming a *McDonalds* world—the gap between rich and poor grows larger each day. Power, privilege, and knowledge are at the root of this devastating destruction and homogenization. But who will take on the environmental challenge—a challenge which means confronting both the enormities of the environmental problems and those responsible for manipulating and controlling knowledge about it.

While I sit pondering this thought the phone rings. The caller is Ron, an environmental activist I had met by chance outside the Botanical Gardens, during an excursion with university ecology students. I can tell by his voice that he is upset. He blurts out almost immediately, "They [the Botanical Gardens staff] started culling the fruit bats last night. While the parents are out finding food, they are knocking the babies out of the trees and giving them lethal injections. Security guards are everywhere and they are refusing to let people in."

It is hard to imagine this scene as only three days ago I had been watching this sociable and affectionate community of fruit bats in the gardens with great delight. The bats are a colony of around four thousand that has returned to roost in the gardens, after being absent from Melbourne's inner city area for over a century. Indigenous to the region, these bats started returning a decade ago. First in small groups, and then in larger numbers, the colony has multiplied and made the Gardens its home. It has become somewhat of a tourist attraction as they chatter on the bare branches in 'Fern Gully,' a reconstructed rainforest. At night they fly out over the suburbs, in search of flowering trees. However, there is a strong campaign to remove them as they denude tree branches for their landing posts, make noise, and soil the ground. But worst of all, according to recent media reporting, fruit bats are 'natural' hosts of a number of deadly viruses, although only one person in Australia (a bat keeper) has been recorded as dying from a bat bite due to a host virus. However, this fear has been implanted in the public's mind. Ron has been trying to build public support to stop the bat culling until the garden staff find more humane ways of controlling bat numbers or an alternative roosting site.

"I am feeling despondent. They have been getting over 300 calls a day from people concerned about the bats, but only four people attended the rally I held at Government House. I am not doing any good. How can I get people to run with the issue, to actually do something?" asked Ron. "Education," I replied. "You are going to have to organize a community education campaign about the alternatives." "Where do I start? How do I do that? I am not a teacher!" he cried. I said, "Ron, what do you think you are doing when you are lobbying for support? You may see yourself as an activist, but standing outside the premier's office is only one strategy. You need more. I think we should make some plans to meet and talk this through."

This chapter explores environmental activism and its potential as a form of environmental adult education. Through an examination of different views or processes of education—formal, non-formal, the behaviourist position and the critical position—it argues that more emphasis needs to be placed on environmental adult education in communities, and that activists should be seen as 'environmental adult educators,' and given the necessary support in their ongoing efforts.

Earth in a Crisis: The Challenge to Education

Education and training on environmental problems are vital to the long term success of environmental policies because they are the only means of mobilising an enlightened and responsible population, and of securing the man [sic] power needed

for practical action programmes. (A statement emerging from the 1972 United
Nations Conference on the Human Environment, cited in Linke 1980, 25.)

Responding to a growing awareness of the environmental crisis, the United
Nations began to develop various programmes such as Man [sic] and the
Biosphere and to organize a number of environmental conferences and
following on these, a variety of environmental education conferences (for
examples, see Tilbury 1994).

As it increasingly became apparent that environmental problems could
no longer be seen as 'scientific' problems (to which science could invent
scientific solutions), but that all sections of the community had to begin to
play a role, UNESCO called for 'new forms of international cooperation'
and the field of 'education to focus its resources on the situation.' In
particular, UNESCO argued that education, and specifically formal and non-
formal environmental education, has an important role to play in
encouraging new visions and practical actions. According to the Belgrade
Charter (UNESCO, 1977), the role of environmental education is to develop
a world population that is aware of, and concerned about, the environment
and its associated problems. The goal is a public with the knowledge, skills,
attitudes, motivations, and commitment to work individually and
collectively toward solutions of current problems, and the prevention of new
ones.

From these early beginnings the field of environmental education had a
formal and non-formal domain. According to United Nations' definitions,
the task for educationists was to respond to the environmental crisis and
develop environmental education curricula, either in a formal or non-formal
context, which would educate individuals to ask the hard questions: Who
took this decision? According to what criteria? With what immediate ends in
mind? (UNESCO 1980). However, the questions of moral decision making
and participatory 'action' caused much debate in education circles.
Contestation over defining the nature and purpose of environmental
education continued into the nineties, with the relationship between
grassroots environmentalism and dominant views of New Right politics
slipping further and further apart. The environmental crisis and radical
social change was reconstructed in more palatable terms of 'sustainable
development.'

Agenda 21 was the 'action' document emerging from the United Nations
Conference on Environment and Development (UNCED), held in Rio de
Janeiro, Brazil, in 1992. The Agenda contained a section on 'education'
based on those definitions that had been developed (and diluted) through the

UNESCO-UNEP (United Nations Environmental Programme) conferences between 1977 and 1987 (UNCED 1992). Paralleling this grand collective of UNCED 'knowledge producers' were small groups of interested environmental educators who collaborated outside the inner sanctum. A more comprehensive definition of environmental education was produced in this way by the Learning for Environmental Action (LEAP) of the International Council of Adult Education (ICAE), during the International Journey for Environmental Education at the International Forum of NGOs and Social Movements, UNCED conference (ICAE 1992). This treaty, called *Environmental Education for Sustainable Societies and Global Responsibility* presented a very different view of environmental education from the definition emerging from *Agenda 21*. With sixteen principles and a strong socially critical orientation advocating social transformation, self-determination, and collective action, the treaty supported the need for both formal and non-formal or environmental adult education, as well as an advocacy role for environmental educators.

Focus on the Formal

During the past twenty years the Australian formal environmental education community has responded to the environmental challenge and developed a comprehensive programme in research, policy development, teacher education, professional development, and school curricula. The guiding principle of much of this work has been the view that children who participate in environmental education programmes during their schooling life will become responsible and active environmental citizens. The emphasis on formal education was supported by a review of the first ten years of the *Australian Journal of Environmental Education* (Andrew and Malone 1995), which revealed that only one of ninety articles published between 1984 and 1994 was concerned directly with non-formal environmental education, even though community involvement has been explicitly stated as an essential purpose.

What has largely not been addressed due to this focus on formal education is the valuable contribution non-formal or environmental adult education can make, and is making, to support environmental actions by local communities (Fien and Trainer 1993). Environmental education was losing its association with the very foundation through which it was established—the growing grassroots environmental movement. Often, the movement actors were perceived as radical activists even though, in light of early definitions of environmental education, they should have been embraced as environmental educators.

Unlike formal education, non-formal environmental adult education is, through definition, specifically community based and therefore has the potential to address the environmental challenge, by responding directly with local environmental issues—at the coalface. For, as the UNESCO (1986, 20) *"Guidelines of the Development of Non-Formal Environmental Education"* aptly state:

> The strength of non-formal environmental education lies in the fact it does not operate within a given set of rules with strict structure, curriculum and examination procedures. Non-formal environmental education is more capable of responding to local environmental issues, which have more social meaning and usefulness to the community and less dominated by academic requirements.

To date the role of educating the community about environmental issues has largely been opportunistic and based on 'information transfer' models where the community have limited opportunities to respond or participate in the educative or decision-making processes. Adult education and community action have been dominated by governmental agencies whose role is to involve community members in specific environmental deeds—clean up Australia Day, curbside pick-ups.

But there are different models of environmental education emerging through the theorizing of critical theory, participatory research processes, and environmental education (Malone 1999a, 1999b). This critical approach to environmental adult education is based on the development of alternative discourses of environmental knowing and acting, enabling the powerless to take action to improve their environment and the quality of their lives.

Behaviourist and Social Transformation Approaches

It is useful at this point to pause and consider the differences between these two approaches to education: education as 'information transfer,' the behaviourist approach, and education as 'knowledge production and information critique,' or the social transformation approach.

As Figure 1 indicates, education as 'information transfer' assumes that an authority or agency is competent in identifying environmental problems and developing solutions drawing on pre-existing bodies of knowledge or empirical evidence—which are thought to be universally applicable (Robottom, et al. 2000, 10). The community becomes the passive 'consumer' of the information and the educative process is characteristic of a top-down learning experience where the 'objects' of the information exchange are designated as instruments for environmental change. Often this approach has an underlying behaviourist orientation—education serves to change the behaviours of the individual or collective.

Figure 1: Two Approaches to Environmental Education

	Information Exchange	**Socially Critical**
View of human nature	Behaviourist	Critical
View of education	Reproducing pre-existing facts	Lifelong process; role of reconstruction
View of environmental education	Knowledge about the environment	Action for the environment; based on principles of environmentalism and social transformation
View of environmental activism	Non-existent	Intrinsically linked; has a symbiotic relationship with the education process
View of knowledge	Pre-ordinate, objective, derived from experts	Generative/emergent; produced by participants in a variety of environmental education settings
Organizing principle of environmental education	Traditional subject disciplines	Environmental issues
Role of texts	Pre-existing primary source of knowledge	Secondary source; critically appraised for worthiness
Educator's role	Authority; knowledge keeper	Collaborative participant and facilitator in the learning process
Learner's role	Passive recipients of knowledge	Active generators of new knowledge; change agents
Community's role	Passive observers of the educational process	Co-learners; active change agents
Role of schooling	Reproduce the status quo	Provide opportunities for school and community members to engage in social change

	Information Exchange	Socially Critical
Role of politics	Apolitical stance	Essential to the learning process
Personal environmental philosophy	Anthropocentric environmentalism	Ecocentric environmentalism

In contrast, education as 'knowledge production and information critique' is based on the assumption that the community has theoretical views, lived experiences, and knowledge about the nature of environmental problems, and that these views, experiences, and knowledges influence and actively shape the development and understanding of what constitutes an environmental problem or issue, and the possibilities for change. An education as 'information critique' perspective engages and demands community members to develop alliances and construct understandings, which, in many instances, are critical of the political persuasion of dominant views. It also endeavours to place the community at the centre rather than at the periphery of the education process. This approach has a socially critical education orientation and is compatible with a symbiotic relationship between environmental education and the environmental movement. It has its roots in the environmental movement and is, therefore, by its nature, oppositional.

Environmental Movement

An oppositional ideology is a pattern of beliefs and values not supported and shared by the majority of individuals within society. An oppositional ideology is defined as an ideology that challenges dominant ideologies by disrupting dominant discourses and providing alternative ways of viewing the world (Malone 1995). Social movements are the construction of a 'fellowship of belief,' a belief held by individuals of an oppositional ideology. Environmentalism as ideology, therefore, is the mortar from which the environmental movement is built.

Building on and using the oppositional ideology, another key element of a social movement is the generation of counter-hegemonic discourses. A counter-hegemonic discourse is an alternative knowledge base through which the movement actors can articulate counter-arguments to dominant knowledge. In the environmental movement, these counter-hegemonic discourses have primarily emerged in two forms: publicly accessible scholarly work, and knowledge produced from lived experience. Firstly, counter-hegemonic discourses exist as knowledge produced by intellectuals who specifically critique dominant scientific knowledge through public and scholarly channels. Authors/scientists such as Rachel Carson (*Silent Spring 1962*), Edward Goldsmith (*Blueprint for Survival 1972*); Vandana Shiva (*Staying Alive*); David Suzuki (*It's a Matter of Survival)* established a core counter-hegemonic discourse. They utilized and constructed this alternative way of reading the environmental crisis to rebuff claims that science had the answers to the crisis. They did this by creating an alternative science—which was just as rigorous as mainstream science—but was written in such a way as to be accessible to the general public. Counter-hegemonic discourses can also be constructed from knowledge produced from the lived experiences of individuals or communities, in particular environmental, historical, political, or cultural contexts. The authors of this knowledge can be varied and diverse (schoolchildren, women, old people, rich, poor) and their writings are often elicited and published through NGOs such as Community Aid Abroad, World Wildlife Fund, Amnesty International, and the International Council for Adult Education.

Another element of a social movement is action. Action in a social movement can take a variety of forms—brief or sustained, institutionalized or disruptive, humdrum or dramatic. Environmental action is often a key tool for both disseminating and developing counter-hegemonic discourses, eliciting wide-spread support from the general public, by making visible environmental issues, and for engaging in specific environmental struggles. Environmental activists are the movement actors, they are the individuals who subscribe to the environmental ideology, engage in a variety of individual and collective actions, and produce and disseminate, through personal and professional means, counter-hegemonic discourses.

Through the following story I further explore the social transformation-oriented approach and counter-hegemonic discourse of environmental adult education and the role of the environmental activist, and how these two very different epistemological constructions of environmental education have been acted out.

Scott's Story

I first met Scott during my doctoral research. I became very interested in the way he was able to verbalize the educational role of his activist life and the strategies he had formulated for encouraging others to do the same. Scott is a parent, a local environmental activist, and resident of a housing estate on the fringes of a large metropolis. He is a self-professed environmental activist who has always had a keen interest in the politics of environmental issues and social justice. He left school when he was fifteen and has been unemployed most of his young adult life. Although having very little formal schooling he has a strong sense of the importance of learning. He believes learning is about being in the world. He applies his experiential knowledge in community re-vegetation projects he is involved in. In this tale he is recalling his involvement in a re-vegetation project in his local community, a project that became highly politicized due to the authorities' apprehension to allow the local residents a forum for expressing their concerns about the toxic condition of the environment. His story is of an environmental activist as environmental adult educator:

> I have always been interested in the environment and have a strong personal commitment to environmentalist philosophies. I believe if we don't change the way we perceive ourselves on the Earth then we don't have a chance. The importance of activist and educational work is very hard for me to qualify. I find myself thinking about it quite a lot and personally the way I see it is that it's more than just an issue of planting trees or bringing back native grasses or cleaning the creek. It is more a social focus—and not so much the task of planting but how that end process is achieved. The process I see is getting people out of their homes, getting together, and getting them to socialize and get some value back into their lives. Being in the Western suburbs we were an easy target. We were easily railroaded because we have no political clout. As far as an identity goes you are always assumed to be off to the factories. No value. No self-respect. No self-esteem.
>
> You need education running alongside environment projects. It is important to run a community environmental adult education programme for the very reason we need to establish a common ideology, some common ground. But you have to try and keep it as informal as possible, because, once people realize it's going to be structured, they shy away. The education just stems from being involved. If people are willing to come in and keep their eyes and ears open, they will start to pick things up.
>
> There has to be a balance between helping the people and helping the environment and, sometimes, you have to start with the people. Get the process working. Get people active before you can start making huge changes to the environment. It is a bit like planting a six-foot already established tree. I said to Peter once, "Why don't you plant six-foot trees on your block?" He said, "A six-foot tree won't grow as fast as a seedling. It's better to put a seedling in from indigenous stock. A six-foot tree, just stop. The growth will be minimal." That's the same with educating people. You learn through experience. You build up knowledge through acting. I am an educator. You are an educator. We are all sharing and learning from each other's experiences.

The main problem we have had is that people don't value their own knowledge or the contribution they have to make to action. A classic example was Norm. He actually said, one night when we were planning for the festival, "Well, who is going to be there on the day?" and I said, "We, were going to be here, us." But Norm said, "No, no, no. What experts are going to be there? Are you going to be there on the day [directed to Mark and Bernice the architects]? Any other experts?"

I take real aversion to that. I didn't go off at him but made a point of telling him, "You're as much an expert as anyone else. More so than Mark because you live here. It gets back to one of Peter's terms, the local providence. That is the beauty of indigenous planting; the seed stock comes from that area. It's specific to that soil, that microclimate and that is the same as the people. If you have been here thirty years, you know the place better than anybody else. Probably that well you do not know it that well. What experts are going to be here? You don't need to have an alphabet after your name before you have something worthwhile to say."

In the end environmental adult education and environmental action is about people, people getting out of their homes and working together for a common goal. We can't do it by talking in committees. We need to have the support of the local people, people who understand and know about the area, what their needs are, what they want, not what the bureaucrats want for us.

I have got a lot out of being involved in this project. More political, more personal in that respect. Just the personal growth, more than anything else. People go to all these assertion classes! Throw them into a community group and they will get it for nothing. And to see how the system works, dealing with councils and agencies. It just demystifies the whole lot. Take that aura away from them being authorities and subjects.

The truth is I don't know any more than anybody else, if anything at all. I believe this project has made a difference, a difference in the way people perceive the role they can take up in their lives. The project is a dedication to the work of the people—a dedication to the community and their struggle to change the balance of power.

We as a community set about to change our situation and in many ways we have, not just the planting but because we as a community hadn't talked at meetings, shared our anger and ideas. If nothing else, at least we can say the community has learnt to identify itself as a community. So next time if we have an environmental concern we know now we can do it and we have the skills to set up a process for change. We aren't scared of making our voices heard.

As this story clearly demonstrates, Scott is an actor in the environmental movement. He has an environmental belief, a personal philosophy, that guides his life. Scott lives his philosophy. Having left school at fifteen he was not well educated in the formal sense. Yet the depth of his commitment to education had emerged from his lived experience as a community activist. Scott's story reveals that the UNESCO view of environmental adult education that emerged in the 1970s about 'mobilizing and enlightening the populus' and 'involving populations in practical action,' is a realistic aspiration.

Conclusion

In this chapter I briefly illuminated aspects of two contested views of environmental education: the information exchange and behaviourist view which see environmental education as changing people's environmental behaviours, and the critical view of environmental adult educators/activists, who see education as a tool for critically appraising dominant constructions of the universal 'truth' and facilitating opportunities for community members to engage in collective environmental actions. In many ways, Scott's story emphasizes this latter view. It demonstrates the role of the environmental adult educator/activist as one to stimulate environmental critique. Critical environmental adult education is not a didactical transmission of knowledge, but, rather, a process for learning through living knowledge which can be translated directly into action.

Environmental adult education has its foundations in the environmental movement and radical education theory, and was developed as a means of advancing, producing, and disseminating knowledge to support environmental action. It contests and constructs alternative worldviews based on the harmonious co-existence of human and non-human nature. Critical environmental adult education is a highly politicized endeavour, and educators who adopt this position become highly politicized environmental actors.

Environmental adult education advocates activism, and that the educator adopt an activist agenda. The activist and the educator in environmental adult education have a symbiotic relationship—an intimate association that is mutually beneficial and often indistinguishable. Environmental adult educators around the world must learn to nurture this educator-activist relationship and reconstruct multiple roles for themselves in their communities.

References

Andrew, J. and Malone, K. 1995. "The First Ten Years: A Review of the Australian Journal of Environmental Education." *Australian Journal of Environmental Education 11*, 131–162.

Carson, Rachel. 1962. *Silent Spring*. Boston: Houghton Mifflin Co.

Fien, J. and Trainer, T. 1993. "Education for Sustainability." In J. Fien (Ed.), *Environmental Education: A Pathway to Sustainability*. Geelong: Deakin University Press.

Goldsmith, Edward. 1972. *Blueprint for Survival*. Boston: Houghton Mifflin.

Gordon, A. and Suzuki, D. 1990. *It's a Matter of Survival*. Toronto: Stoddart Publishing Co., Limited.

Malone, K. 1995. Celebrating Our Subjectivity: Research as Lived Experience. *South African Journal of environmental Education 14*, 10–14.

Malone, K. 1999a. Reclaiming Silenced Voices Through Practices of Education and Environmental Popular Knowledge Production. *Canadian Journal of Environmental Education 4*, Summer, 231–242.

Malone, K. 1999b. Environmental Education Researchers as Environmental Activists. *Environmental Education Research 5*, 2, 163–177.

Robottom, I. Malone, K., and Walker, R. 2000. *Case Studies in Environmental Education: Policy and Practice*. Geelong: Deakin University Press.

Shiva, V. 1989. *Staying Alive: Women, Ecology and Survival in India*. London: Zed Books.

Tilbury, D. 1994. "The International Development of Environmental Education: A Basis for a Teacher Education Model?" *Environmental Education Research 13*, 2, 1–20.

UNESCO. 1977. "Trends in Environmental Education." Paris: UNESCO.

UNESCO. 1980. "Environmental Education in Light of Tbilisi, Presses." Paris: Universitaires de France.

UNESCO. 1986. "Guidelines of the Development of Non-Formal Environmental Education." Environmental Education Series No. 23. Paris: UNESCO-UNEP International Environmental Education Programme.

Chapter 10

Environmental Adult Education Within Participatory Research: Learning at the Edge of Social Movements or Cultural and Ecological Sustainability

Jan Woodhouse
Northern Illinois University, USA

Introduction

Most of the work that I have done over the last thirty years has been in the continental United States, and from the social location of a white, lower- to middle-class woman whose ethnic roots were in the Midwestern agricultural traditions.

I began my professional life with a bachelor's degree in elementary education. My first position was as a teacher of grade 1 students in a rural farming community in central Illinois, 15 miles from where I was born and raised. Subsequently, I worked in a suburban Chicago school teaching all subjects to grade 5. A year later we departmentalized, and I taught only science to grade 5 students. This was during the first wave of the ecology movement in the United States. This work led me to pursue a master's degree in outdoor/environmental education. I began to understand systems ecology and how to do the work to bring about ecological sustainability.

Following my husband to the southwest so that he could work on his doctorate, I took a position with a Catholic school located in the barrio of Flagstaff, Arizona. This was a kind of Peace Corps experience. The people were very poor and socially marginalized. Educational resources were limited, and the staff was young and transient. I had to be creative, persevering, and use every skill and knowledge base I had ever been exposed to in order to make things happen. I taught grade 4, served as assistant principal, and developed an environmental education curriculum for grades 7 and 8.

My husband's dissatisfaction with the Ph.D. program at Northern Arizona University motivated us to transfer his studies to Arizona State University (ASU) in Phoenix/Tempe. As he proceeded with doctoral work in plant physiology, I supported us as assistant director of a women's residence hall. I was also invited by the Dean of Students to be part of a team to write a grant to establish a course for women returning to education. This was the mid-seventies, and women's studies programs were just beginning to be explored for empowerment and liberating potential. Our grant was funded and we team-taught the course at an off-campus site. I also began working with Phoenix Parks and Recreation to design weekend workshops for teachers, agency personnel, and families. We brought the background into focus, teaching about the natural and cultural history of local urban and regional parks.

After one year in residence hall management, I took a position with the Physics Department at Arizona State University as a graduate student in natural sciences, and for the next five years helped coordinate and instruct teacher-training programs around the state in outdoor/environmental education, energy education, and general sciencing. These programs were diverse in terms of geography, race, and class: Some were in the affluent suburbs of the Phoenix Metro area; some were in the inner city; some in the rural highlands; and some on the Indian Reservations. I began to realize that my formal education had really taught me two essential skills: how to go into an environment and assess what there was to work with, in terms of human resources and physical resources; and to set up a structure of information, experience, and support to achieve certain goals. These skills were what enabled me to work effectively in a variety of contexts and to make participatory action research and participatory community development a preferred way of working in the future. I also began to realize that I had an innate propensity to look ecologically to see the "big picture" and to be able to understand how the elements in that picture, the human and non-human, were connected. I was much more skilled at working on that level than with the details or the micro-level. I was also much happier.

At that same time, as the women's movement re-emerged, I began to experience and understand the ways that women were oppressed and to read and study broadly on women's issues (including my own co-dependency and subordination). Each job opened opportunities for the next, and over the years I worked in long- and short-term endeavours, public and private, paid and volunteer, including but not limited to: coordinating curriculum development projects; conducting leadership and supervisory training for business and industry; founding and coordinating rural women's support groups and services; founding and coordinating local and regional tourism development efforts; co-founding and coordinating grassroots watershed education and monitoring programs; publishing a bioregional magazine; and coordinating adult resources and student leadership development for a community college. I also made several houses into homes and helped raise (at least) four children to adulthood.

Like many women of my generation, I didn't have the luxury of a carefully constructed career path like many of my male counterparts. We were usually in a place because of someone else's needs and work, husband or family. We 'made do.' We began where we were, used what we had, and did what we could. Then, we were in a new place to begin again. Line by line we composed a life. Most of what I did, and what women like me did, was similar to the women I have written about in this chapter. Much of my work evolved because I saw that something was wanted and needed. I had no conventional power to bring to the situation (and in some places, my formal education was not considered an asset, either). I did not have the power of position or money to bring about change. I had to work in different ways. The focus was on the problem, not on self-promotion or empire building. I/we worked unselfconsciously, acquiring the knowledge and skills we needed as we worked, and learning creative problem solving and perseverance as we overcame entrenched resistance to change. In retrospect, it is clear that we were part of a global social movement and practicing participatory community development. We didn't call it that, then, and we had little awareness of this greater context. We were just doing what needed to be done, which in some cases was simply surviving, physically and psychologically.

It is only now, as I have returned to a doctoral program in adult education, that I have the advantage of distance, time, and counsel to critically reflect on the work that I have done. This article is a part of that process. Having worked broadly for so many years, I began to recognize patterns to the ways that women work to improve the quality of their lives and the lives of the people they care about. I also began to see the interface of social movements, environmental adult education, and participatory approaches to change. I have been a part of each of these paradigms. I understand how they relate to and support each other. I also recognize that in many places the people involved in these processes do not see this interface.

Adult education in the United States is only beginning to acknowledge and include examination of environmental issues as an imperative to cultural sustainability. Problematically, environmental education discourses are almost solely about elementary and secondary school systems, leaving community-based programs, higher education, and social movements at the margins. Participatory research has also not been discussed as an environmental education for adults. It has its own language. Because participatory research focusses on a particular context, it may fail to see the bigger picture: the social movement paradigm within which this context functions. Academia exacerbates this myopia because it forces the researcher to work in-depth with one small piece of the puzzle. I believe that this approach is necessary to knowledge formation, but I also believe that, at some point, someone has to put the pieces together. I believe that this is more critical today than ever. Everything has always been connected to everything else, but the forces of globalization cause what happens in one place on the planet to have implications to other places, in profound ways.

While all of these stories begin with the struggles of people in a specific place, the forces that create the problems and/or the forces that are brought to play to solve the problems cross the borders of race, class, disciplines, political organization, power bases, and geography. They demonstrate the border crossings that are endemic to these efforts. The projects are also different in many ways and therefore the analysis is not parallel. Considered collectively, they demonstrate the power and potential of local efforts to bring about cultural and ecological sustainability.

I discovered the story of Madha Patkar and the Narmada Valley Project while doing research on women, education, and environment in India. In 1997, I participated in an adult education study tour to India. I met many women who were working at the local, regional, and national level in programs to increase women's agency. Patkar's story is exemplary of how one woman's efforts can start a social movement. It is also exemplary of the personal and political struggle that may result from what starts as a simple effort to address what is wanted and needed. It illustrates the connection between the local and global.

Louise Fortmann's story demonstrates how a woman from a white, urban, middle-class culture in the United States can facilitate community empowerment in a black, rural, lower-class culture in Africa when participatory research is carried out effectively. I discovered her story last year during a course in participatory action research. Her work is also an example of un-named environmental adult education.

The work of Darlene Clover and Shirley Follen was uncovered during a course in international development. Their book, *The Nature of Transformation*, was referenced in one of the course readings. I knew immediately that we were working from a similar line of thinking. I was able to contact Darlene, set up a meeting with her, and have subsequently worked with her and Shirley at a popular education conference in Tucson, Arizona, and am working with Darlene on creating an International Institute in Environmental Adult Education.

As I study environmental adult education within participatory research and social movements, I am inspired and encouraged about the future of the planet. I could have filled this book with case studies from every continent. The discouraging fact is that most of these efforts go unacknowledged beyond the local or regional level. Yet, they are occurring in such a pervasive and persistent manner that they have the power to transform civil society, and, therefore, ultimately, to moderate the impact of globalization and other destabilizing forces. These stories must be told. To that end, I offer this chapter.

Rationale and Structure

The critical question today is not: Is there an environmental problem? The question is: How long can the Earth's biosphere support life as we know it? (Bowers 1995; Brown et al. 1996; Carson 1962; Commoner 1971; Ehrlich and Ehrlich 1991; Gore 1993; Orr 1992 and 1994; Postel 1996; Suzuki 1997; Tilbury 1995). Subsequently, we must ask: How do we construct or re-construct communities so that ecological and cultural sustainability is the foundation of change? What is the role of education and research in bringing about this change? The knowledge needed to construct systems (local to global) that can effectively respond to these questions will be found in a spectrum of contexts, and in understanding the ecology of the place and its problems in the broadest application of that concept (Bowers 1995 and 1993; Clover et al. 1998; Hicks and Holden 1995; Orr 1994; Palmer et al. 1998; Tilbury 1995).

This chapter examines the characteristics of environmental adult education that are interdependent with selected participatory action research (PAR) projects. These projects are profiled because they represent the work that is being carried out around the world, through grassroots initiatives to preserve local ecologies and cultures. Since, in that context, environmental adult education becomes an agent of change, a discussion of the relationship of these initiatives to social movements is included. In other words, unlike typical academic endeavours that examine only one piece of the problem, this discussion will look, practically and theoretically, at what happens at the interface of environmental adult education, participatory research, and social movements. I suggest that examining this interface is critical to understanding the imperative of local initiatives to bring about cultural and ecological sustainability. Three projects will be profiled which manifest elements of the practical and theoretical foundations of these claims.

The Narmada Valley Project, India

Madha Patkar came to the Narmada Valley of northern India in 1985 to study the impact of the proposed construction of the Sardar Sarovar Dam. The construction of the dam would submerge local villages and would mean a complete reordering of the lives of the people who claimed those places as home. Patkar was a social worker, researcher, and activist. Her research was carried out over the next few years and involved extensive travel by foot, bus, and boat throughout the territories in the submergence zone. She lived with the people, learned about their culture and ecology, and listened to their concerns and fears about the future.

The Sardar Sarovar Dam was part of a larger development project of the World Bank that ultimately called for the construction of thirty major dams on the Narmada River and its tributaries. Despite World Bank and engineers' claims to the contrary, Patkar anticipated the inevitable and potential eco-catastrophes and social disruption that this project would cause. Ninety-two thousand acres of fertile fields and forests would be submerged. One hundred thousand tribal peoples in 245 villages would lose all or most of their land. The lands and lives of thousands more would be impacted. While the project plans called for these people to be 'resettled and rehabilitated,' Patkar knew this meant that most of them would end up living in urban slums or working as migrant labourers in the sugar cane fields.

Her work developed trust among the villagers and attracted the attention and commitment of young activists from outside the valley. Throughout the region, efforts were initiated to organize and educate the people. This effort was eventually a catalyst to organize the country's environmental groups. Patkar became a key leader in raising social justice issues such as "Who pays and who benefits?" Finally, after testimonies before the U.S. Congress and the World Bank, the World Bank stopped its plans to fund one of the large dams. The struggle to protect indigenous cultures and to maintain the ecological stability of the region continues. But increasing pressure from human rights and economic forces is slowing dam construction and causing re-evaluation of many projects. The Japanese government has withdrawn its funding, and members of the European community have urged the World Bank to stop financing the project altogether (Wallace 1996).

Gender and Trees in Zimbabwe, Africa

Louise Fortmann, Professor of Natural Resources Sociology at the University of California, Berkeley, explored the interface of three issues: gender, tree tenure, and tree planting, in two villages located in two different ecological zones in Zimbabwe, Africa. This research grew out of the belief of feminist political ecologists that gender plays a role in how natural resources are used and managed. Fortmann (1996) hoped to obtain data to address these issues and to explore how participatory methodologies might facilitate the gathering of the data, generate knowledge, and result in the sharing of expertise. She hired seven villagers to work as a research team assisting with the data collection. She hoped these community members would develop a consciousness about their problems, would become experts and spokespersons on community problems, and would develop a support and action network. Her hopes were realized. For example, after the project was completed, five of the women, all welfare mothers and none with a high school diploma, organized a tour of the county for the dean of the College of Agriculture, University of Zimbabwe, to discuss and show him the extent of rural poverty.

Fortmann's research took five basic forms, four of which employed an environmental adult education component:

1   Resource mapping—Groups of men and women were asked to draw (with a stick in the soil) a map of where they found their tree products.

2   A questionnaire survey—Fortmann asked people about what trees they used and for what purpose. With the help of the village research team, she collected physical specimens of most of the 122 indigenous trees listed, grouped the trees by habitat, and cross-referenced the list with the work of botanist Robert Drummond. The

results of this effort were published, acknowledging the research team as senior authors of *The Use of Indigenous Trees in Mhondoro District.*

3   Wealth rankings—Fortmann worked with the local research team to create a 6-point scale that would indicate what rich people had and what poor people had. Each village researcher ranked respondents who chose to be included.

4   Public presentation of the research—The research team was asked if they wanted to present their findings. They were coached to organize their presentations and to speak publicly. The result was a villagewide event that included speeches and an eloquent prayer about trees.

Fortmann learned a variety of lessons from this project. She found that the resource mapping exercise had to be done with men and women separately, otherwise women often submitted to the knowledge of the men and were less likely to make their voices heard. This exercise also showed that men and women in this community used the same spaces differently, they used different spaces, and that the knowledge about trees was highly gendered. Finally, she concluded that participatory methods built community through the research process.

Her research contributes to the more generalized claims that participatory research can be used to collect and analyze data about various phenomena to increase knowledge for both the academy and the people in the place where the data is gathered. The report also confirms ecofeminist claims that women's knowledge and use of resources is different than those of men. In many cases, women's knowledge of local resources is more extensive than men's because women are the members of the community responsible for the gathering, managing, and using local resources.

This project conformed to standard PAR principles, in that it was designed to utilize the knowledge of the local people as much as possible, to benefit from their skills, and to teach them new skills, and to leave something behind that would be of value to the participants. It is not, however, clear from her reports how that knowledge might be used in the future. Fortmann's research deviates from the participatory model in that neither the idea for the research nor the research methods came from the participants.

The International Journey for Environmental Adult Education

A longer-term project used environmental adult education as the focal point of the research. Between 1994 and 1997, Darlene Clover, senior researcher with the Transformative Learning Centre in Toronto, Canada, and Shirley Follen, adult educator and workshop facilitator, worked with communities and organizations across Canada and in other parts of the world. They developed environmental adult education workshops using participatory action research designed to: begin from the people's own potential to address and find solutions to their own problems. These workshops incorporated a variety of strategies to accomplish that objective. Activities facilitated participants to make links between environmental and other social issues, make links between the local and the global, encouraged reflection and personal and collective evaluation, and promoted concrete actions. Clover and Follen also tried to make the workshop experiential, active, and fun. A more detailed analysis of the elements they found to be essential in the success of these workshops can be found in the book that resulted from this work (Clover et al. 2000).

These workshops used the knowledge and resources of the participants in many of the same ways that were described in the previous projects. Clover and Follen's work differed from the other projects in that the length of time working with each group was shorter. Both the participants and the researchers benefitted. The researchers succeeded in their objectives to identify and field-test effective educational approaches and strategies for community-based environmental groups, adult or popular educators, municipal organizations, activists and organizers, and other people interested in the future of their communities and the planet. Clover and Follen also examined the factors of resistance and of women's voices in the participatory process.

Clover and Follen are candid about what they learned because of what didn't work and because of what was not anticipated. For example, they learned to make last-minute adaptations when the participants were 'radically different' from what they expected; when many more or fewer people showed up than what was planned for; when elderly participants' stories of nature were 'frightening and horrible experiences,' and when the social, economic, and educational experience of the participants was so radically diverse. They reflect that their greatest challenge was to "re-frame or re-shape many years of community-based educational methodologies and principles from an ecological point of view" (Clover et al. 2000, 1).

What are the Characteristics of the Environmental Adult Education That Take Place Within Participatory Action Research Projects?

Environmental adult education that evolves within participatory action research projects proceeds from the belief that knowledge is created, or indigenous knowledge is applied, to increase self-reliance. People re-connect with the natural word and develop a deeper understanding of home and place by using nature and the community as teachers and sites for learning. Through a variety of place-based activities, people learn to take back control of their environment.

The resolution of problems that threaten cultural and ecological sustainability is being negotiated through such efforts and through social movements. These movements usually begin as local initiatives: rural, loosely organized, unfunded, generated from and dependent on the actions of those affected by the problem. They may evolve to be regional, national, or multi-national in scope, technologically based, funded by government and/or corporate interests, and administrated by specially trained persons who probably aren't directly or immediately affected by the problem (Woodhouse 1998). These movements involve adult education as an agent of change on a multitude of levels. Participatory action research is increasing as a means to organizing local knowledge, skills, and resources in change efforts. Rocheleau, Thomas-Slayter, and Wangari (1996) present an excellent study of the global scale of these endeavours.

How are PAR Principles Being Applied to These Efforts?

The projects discussed in this chapter reflect the attributes of environmental adult education. They also require a spectrum of interdisciplinary effort; however, that dimension of the enterprise, while apparent in the projects examined, is not a focus of this paper. These projects also exemplify the following principles of PAR:

1   *PAR is place-based, meaning it evolves out of the "place" of the people and problem(s) being addressed.* Using the geography, sociology, and dynamics of the place (rather than constructing and imposing a problem or process that only benefits the researcher), facilitates a research process that empowers rather than exploits the participants. The participant becomes subject rather than object.

2   *PAR invites a deliberate and systematic empowerment of the people with whom the research is being carried out.* Activities increase both the knowledge and skills base of the participants/community. The local knowledge of self-designated leaders and others are used to refine the ideas that the researcher brings to the project, so that the methods used are appropriate and effective (and the least intrusive and exploitive) for the people and problems addressed.

3   *PAR considers the impact on the community of the publicity that the research may elicit.* Publishing the results of research may bring more researchers, TV cameras, and other media forces to the community. The potential for, and the impact on the village of this publicity should be considered before the PAR is initiated. The people should understand that this potential exists and should agree to accept this possibility. Appropriate training or structures should be set in place so that local forces can manage the impact of this publicity. In the best case scenarios, this preparation can be part of the empowering process, and can enable the local people to benefit from the publicity—to manage it in such a way that their needs are served.

4   *PAR is flexible and adaptive.* It may begin with a plan, but remains open to changing the plan to accommodate local and changing dynamics. This requires an attitude of openness and a process that includes effective communication strategies to make sure that people are able to continually express their ideas about what is unfolding. This also implies a style of 'leadership' or 'facilitatorship' that accepts that 'none of us is as smart as all of us' and that all have a right to participate in defining problems and making decisions.

5   *PAR examines the gendered implications of issues.* In most cultures, there is or may be gender-related implications of resource use, sustainable practices, and other cultural dynamics. The beginning stages of responsible PAR include a means of assessing how gender effects community structures, and constructs the project to reflect that understanding.

6   *PAR considers that 'everything is connected to everything else.'* The dynamics of a community (or a problem) represent an ecology—an interrelationship among the living and non-living elements of that community. This is similar to understanding the 'context' of a problem, but goes beyond context. Illuminations of context identify

all the players but may not always clarify and verify the relationships among the players. And, it may not consider the synergy that is manifested by these interrelationships.

7   *PAR is based on the people's own categories of knowing (what is considered knowledge) and language (how they talk about what they know).* This is essential for effective communication of all kinds, particularly that which involves gathering reliable data. Applying this principle will ensure that the participatory methods serve both the researcher and the researched. It will also ensure that the results of the PAR are comprehended and owned by the local people.

8   *PAR acknowledges that there is a spectrum of approaches that may be used to extract and construct the knowledge and data.* It does not privilege any particular approach, but rather looks at what is wanted and needed and applies or creates approaches that will meet that challenge. These approaches, therefore, may be interdisciplinary, may involve a collective of expertise from within the community or without, and may utilize both qualitative and quantitative collection and analysis of data.

9   *PAR gives credit where credit is due.* PAR acknowledges the participation and contributions of the people who participate. Through this acknowledgement, PAR supports the local expertise and agency of the people involved.

10 *PAR leaves something useful behind.* The process will not only produce useful data, but, also, it will involve the participants in using skills and techniques that they can learn from and replicate in the future—without an 'outside' facilitator. Participatory action ensures that the people and the place within which the work is carried out benefit from the research.

The Theoretical Basis of the Greater Context

The previous discussions have illuminated the ways in which adult education is a part of environmental adult education, participatory action research, and social movements. Understanding that interface or dynamic is less complex if we look at their common theoretical basis. I suggest that this is predominantly the relationship that exists between the content and structure of the adult education enterprise within the social context. I will define content to mean the focus or intent of the enterprise, including the knowledge base(s) that the enterprise will utilize. I will define structure as the plan and/or process of the enterprise, including a definition and agreement on the ground rules or covenants for the engagement in the enterprise   that is, who has got the power to do what, when, how, where, and to, or with, whom. And, I will define social context to include the entirety of the relationships between the human and non-human landscape: the ecology of the enterprise.

In all three paradigms (environmental adult education, participatory research, and social movements), the design of the adult education enterprise begins with an understanding of the social context of the adult learner and the context within which the enterprise is going to take place—this may be two different contexts. The second step is to define the content: that the adult learner believes is wanted and needed. Then, the structure can be determined; the process used to carry out the enterprise should be constructed to be compatible with the social context and complementary to the content. This approach can be effective in any social context when applied with an attitude of mutual understanding, acceptance, and respect. This approach is consistent theoretically with Habermas' analysis of communicative action and emancipatory interests (Collins 1995, 29–30), even if some critics would argue that the inclusion of this analysis has not made much headway into adult education (Collins 1995, 109).

Historically, even adult education programs in the developing countries have been a result of what Young called the "unholy trinity" of western missions, colonial administrations, and multinational corporations (Ewert 1989, 85). These efforts had, as their organizing principle and mission, the interests of the dominant culture, not the people of the countries to which the assistance was being 'given.' Today, as a result of the work of Freire, UNESCO, and many others (Ewert 1989), this approach is changing to models that reflect the principles discussed previously, and which, therefore, engage a more participatory and environmentally sensitive political process. These approaches are bottom-up (from the grassroots/from the needs, knowledge, agency of the people) instead of top-down (from institutions of authority). They are multi-dimensional (Hamilton and Cunningham 1989, 447), such as formal adult education programs, informal/popular education programs, reorganization of institutional structures, redefinition of roles, redefinition of goals, etc. They are empowering processes intended to dismantle colonial systems; for example, programs which incorporate critical consciousness, education for mobilization, popular education, and integrated community-based development (Ewert 1989).

These approaches also collaborate Matthias Finger's conception of the impact of the new social movements on adult education, looking at shifts in the intent of the aims of the enterprise and the place of transformation. He claims that what happens to the adult in this enterprise is mainly informal, local, and communitarian, based on concern, commitment, and experience, rooted in and contributing to the development of a local culture (Finger 1989, 19). The new spaces created through this struggle may be a result of, and will certainly allow, the border crossings that Giroux characterizes as indicative of effective social movements (Giroux 1992).

Understanding the theoretical basis of the interface of PAR, environmental adult education, and social movements requires a cross-disciplinary examination of sources. The history of the adult education enterprise can be traced from the liberal theories through the progressives, and to the human growth potential traditions (Beder 1989). Most learning at the edge of social movements today is a variation on these themes (Savignon 1993). This is evident in Rocheleau, Thomas-Slayter, and Wangari's recent text, *Feminist Political Ecology: Global Issues and Local Experiences*, mentioned earlier, as it presents a most comprehensive review of the global scale of social movements, including PAR for ecological and cultural sustainability (Rocheleau et al. 1996).

Conclusion

The projects selected for examination in this chapter manifest characteristics exemplary of PAR that interface with environmental adult education: the most successful efforts at resolving acute environmental problems evolve from a context, begin locally, and are limited in scope. The intention of the action taken, and, therefore, often the focus of the participatory research, is to change or block a force that threatens something of value to the person(s) and/or community. The learning that takes place for that to happen is initially an ancillary activity. Women are at the centre of these political, educational, and research efforts for ecological and cultural sustainability. Most of the women (and men) in these projects have to learn skills to overcome problems they did not create. Action precedes education. Despite all the religious, legal, and ancient cultural traditions that restrict women's agency in some parts of the world, when there is a crisis, they act. They organize and act, or they act independently and then they organize. The crisis gives them permission to ignore convention. Once the initial crisis has been resolved, participants stay involved to address the problem(s) that provoked the crisis. At this point the need for more education may be recognized. Formal and informal education becomes interdependent with action as the individuals and/or systems continue to develop. Local efforts become catalysts for broader-based efforts. 'Broader-based' may be defined in terms of the scope of the issues or the geography. In other words, this is the point at which local initiatives evolve into social movements.

The case studies summarized here are a microcosm of what is going on around the world. One chapter and one woman's synthesis can only begin to examine the imperatives that seem to be inherent in the interface of environmental adult education, PAR, and social movements. Based on personal experience and on an examination of the literature, a reasonable structuring of this phenomenon has been outlined. A kaleidoscope of the theoretical foundations that are common to the three paradigms has been offered. I hope that this discussion will invite more practitioners and theorists to explore and expand on these considerations; that it will illuminate the practice and theory of adult education as an agent of change as it is manifested in these three paradigms; and that it will help move the consideration of adult education as an agent of change from the margins to the centre of contemporary discourse.

References

Beder, H. 1989. "Purposes and Philosophies of Adult Education." In S.B. Merriam and P.M. Cunningham (Eds.), *Handbook of Adult and Continuing Education.* San Francisco: Jossey-Bass Inc., 37–50.

Bowers, C.A. 1993. *Education, Cultural Myths, and the Ecological Crisis: Toward Deep Changes.* Albany, NY: State University of New York Press.

Bowers, C.A. 1995. *Educating for an Ecologically Sustainable Culture.* Albany, NY: State University of New York Press.

Brown, L., Flavin, C., and Postel, S. 1996. "A Planet in Jeopardy." In R.M. Jackson (Ed.), *Global Issues 96/97.* Guilford, CT: Dushkin Publishing Group/Brown and Benchmark Publishers, 98–100.

Carson, R. 1962. *Silent Spring.* Boston: Houghton Mifflin.

Clover, D.E., Follen, S., and Hall, B. 1998. *The Nature of Transformative Environmental Adult and Popular Education.* Toronto, ON: University of Toronto Press Inc.

Collins, M. 1995. *Adult Education as Vocation: A Critical Role for the Adult Educator.* New York: Routledge.

Commoner, B. 1971. *The Closing Circle.* New York: Knopf.

Ehrlich, P.R. and Ehrlich, A.H. 1991. *Healing the Planet: Strategies for Resolving the Environmental Crisis.* Reading, MA: Addison-Wesley Publishing Company, Inc.

Ewert, D.M. 1989. "Adult Education and International Development." In S.B. Merriam and P.M. Cunningham (Eds.), *Handbook of Adult and Continuing Education.* San Francisco: Jossey-Bass, Inc., 84–98.

Finger, M. 1989. "New Social Movements and Their Implications for Adult Education." *Adult Education Quarterly 40,* 1, 15–22.

Fortmann, L. 1996. "Gendered Knowledge: Rights and Space in Two Zimbabwe Villages." In D. Rocheleau, B. Thomas-Slayter, and E. Wangari (Eds.), *Feminist Political Ecology: Global Issues and Local Experiences.* London: Routledge, 211–223.

Giroux, H.A. 1992. *Border Crossings: Cultural Workers and the Politics of Education.* New York: Routledge.

Gore, A. 1993. *Earth in Balance: Ecology and the Human Spirit.* New York: Houghton

Hamilton, E. and Cunningham, P.M. 1989. "Community-Based Adult Education." In S.B. Merriam and P.M. Cunningham (Eds.), *Handbook of Adult and Continuing Education.* San Francisco: Jossey-Bass, Inc., 439–450.

Hicks, D. and Holden, C. 1995. "Exploring the Future: A Missing Dimension in Environmental Education." *Environmental Education Research 1,* 2, 185–193.

Orr, D.W. 1992. *Ecological Literacy: Education and the Transition to a Postmodern World.* Albany, NY: State University of New York Press.

Orr, D.W. 1994. *Earth in Mind: On Education, Environment, and the Human Prospect.* New York: Island Press.

Palmer, J., Suggate, J., Bajd, B., Hart, P., Ho, R., Ofwono-orecho, J., Peries, M., Robottom, I., Tsaliki, E., and Christie, V. 1998. "An Overview of Significant Influences and Formative Experiences on the Development of Adults' Awareness in Nine Countries." *Environmental Education Research 4,* 4, 445-464.

Postel, S. 1996. "Facing a Future of Water Scarcity." In R.M. Jackson (Ed.), *Global Issues 96/97.* Guilford, CT: Dushkin Publishing Group/Brown and Benchmark Publishers, 162–164.

Rocheleau, D., Thomas-Slayter, B., and Wangari, E. (Eds.) 1996. *Feminist Political Ecology: Global Issues and Local Experiences.* London: Routledge.

Savignon, S.J. 1993. "Communicative Language Teaching: State of the Art." In S. Silberstein (Ed.), *State of the Art TESOL Essays.* Alexandria, VA: Teachers of English to Speakers of Other Languages, Inc.

Suzuki, D. 1997. *The Sacred Balance: Rediscovering Our Place in Nature.* New York: Prometheus Books.

Tilbury, D. 1995. "Environmental Education for Sustainability: Defining the New Focus of Environmental Education in the 1990s." *Environmental Education Research 2,* 195–212.

Wallace, A. (Ed.) 1993. *Ecoheroes: Twelve Tales of Environmental Victory.* San Francisco: Mercury House.

Woodhouse, J.L. 1998. *An Examination of Experiential Approaches to Oral Skills Development.* Unpublished manuscript.

http://www.narmada.org/sardarsarovar.html retrieved: April 4, 2002
    Following a writ petition by the NBA calling for a comprehensive review of the project to take into consideration all the concerns raised, the Supreme Court of India halted construction of the dam in 1995 at a height of 80.3m. However, in an interim order in

February 1999, the Supreme Court gave the go ahead for the dam's height to be raised to a height of 88m (85m + 3m of "humps"). The resultant increased flooding in the monsoon season of 1999 can potentially drown the homes and lands of as many as 2000 tribal families in about 50 villages.

On October 18, 2000, the Supreme Court of India delivered its judgement on the Sardar Sarovar Project. In a 2-to-1 majority judgement, it allowed immediate construction on the dam up to a height of 90m. Further, the judgement authorized construction up to the originally planned height of 138m in 5-meter increments subject to receiving approval from the Relief and Rehabilitation Subgroup of the Narmada Control Authority. It should be noted that the court has said nothing new on the matter. The Narmada Water Disputes Tribunal Award states that land should be made available to the oustees at least an year in advance before submergence [Clause IX, Subclause IV(2)(iv) and Subclause IV(6)(i)]. The essentially unfettered clearance from the Supreme Court has come despite major unresolved issues on resettlement, the environment, and the project's costs and benefits.

http://www.biology.lsa.umich.edu/courses/nre270/NGO%20activists/Medha%20Patkar.html

As an outgrowth of her work to stop dam construction, Patkar has helped establish a network of activists across the country—the National Alliance of People's Movements.

http://www.goldmanprize.org/prize/prize.html

Madhu: 1992 Winner/Asia; The Goldman Environmental Prize is given each year to six environmental heroes—one from each of six continental regions: Africa, Asia, Europe, Island Nations, North America, and South/Central America. Initially each recipient received a $60,000 award from the Goldman Environmental Foundation. The award stipend has been raised three times since and currently stands at $125,000.

Chapter 11

Towards Transformative Environmental Adult Education: Lessons from Global Social Movement Contexts

Budd L. Hall
University of Victoria, Canada

This chapter draws on the previously unpublished results of the 1993–1994 Transformative Learning Through Environmental Action Project (OISE, 1994) which examined the learning dimension of six environmental action campaigns in six different countries around the world, all of which focussed on aspects food security, preservation, or consumption. These results are re-examined in the light of the goals of this book as a contribution to the growing interest in the theory and practice of environmental adult education. In addition to the background and objectives of the original study, the principles of transformative learning that were derived, practices and processes and indicators of success are explored.

Background

The early twenty-first century has seen a dramatic increase in the visibility of social movement and civil society actions through the new style confrontations, that is, at the World Trade Organizations Meeting in Seattle, the International Monetary Fund meetings in Washington D.C., and the Organization of American States meetings in Windsor, Canada. These well-organized and sophisticated interventions have heightened public and media attention to the growing role and activities of both local and global social movements and structures of civil society. Within the scholarly community of adult and lifelong learning, there has been a parallel growing interest in the relationship of learning to social movements (Welton 1993; Holford 1995; Mojab 2000) and the role of adult learning within global civil society contexts (Hall 2000; Tandon 2000). This chapter is both a contribution to our understanding of environmental adult education and to the contemporary interest in the strategic role of social movements and global civil society formations. This chapter is based on data on the Transformative Learning Through Environmental Action Project that has, until now, not been available in book or journal form. This project was undertaken between 1992 and 1994 by the Transformative Learning Centre at the Ontario Institute for Studies in Education (OISE) or the University of Toronto, the Faculty of Environmental Studies of York University, and Communication, Information and Education on Gender (CEMINA), a Brazilian-based environmental NGO. This large scale comparative and international research project was funded by the International Development Research Centre of Canada to investigate ways in which learning emerged, and was stimulated and supported in different environmental social movement contexts around the world. These contexts were in Brazil, Canada, El Salvador, Germany, India, Sudan, and Venezuela. Coordinated by Moema Viezzer in Brazil, Darlene Clover, Budd Hall, Edmund Sullivan, the late Dian Marino and Leesa Fawcett in Canada, the project developed as a contribution to, and a way of following up, the adult education dimensions of the Earth Summit in Rio de Janeiro in June of 1992.

Several years of networking in the international adult education community, especially in preparation for the Earth Summit, raised a number of questions:

1   How could the learning dimension of the environmental movement be strengthened?
2   What can be learned from social movement environmental action campaigns about the ways in which learning takes place and can be most enhanced?
3   Which combinations of pedagogical practice hold out the most promise for transforming relations of power and perception?

The study team chose the term 'transformative learning' because it is a major element of environmental adult education and other critical theories such as popular education and feminist adult education. Moreover, its very name resonates a desire for change. Transformative learning, as it was framed in this study, referred to the process of learning, whether in formal or non-formal education settings which is linked to changing the root causes of environmental destruction or damage. This includes changes in relations of power, gender relations, and other patterns which allow for a healthy relationship with the Earth (OISE 1994).

Objectives of the Study

The objectives of the study were to:

1   identify indicators of success for transformative learning within environmental action contexts;
2   undertake an international survey of transformative education initiatives;
3   develop a number of conceptual working papers and case studies dealing with concepts of transformative learning through environmental action; and
4   organize a collaborative workshop to analyze how transformative learning works.

Methodology

The study was a participatory and collaborative effort by the teams at the three sponsoring organizations which brought diverse approaches and experiences in partnership with a team of scholar-activists who were responsible for writing the case study reports. For example, OISE/UT had extensive experience in adult and popular education theory; the Faculty of Environmental Studies at York University with environmental education, nature and society; while CEMINA brought experience in feminist environmental popular education.

Early in the design of the study, it was agreed that the range of experiences of transformative learning in the field of environmental action was too large and diverse to be able to provide enough points of comparison for one study. As a result, it was agreed that the case studies would be drawn from experiences of environmental action in the context of food production, distribution, and consumption. Food is life itself and all social economic and political relations with nature can be understood from the point of view of food, or even, as we were to discover in the study of the Navdanya (nine seeds) project in India, from the point of view of the seed. As Leesa Fawcett (1993, 5) noted in the final report of the study, "Everything we put into our food, we eventually eat."

The Case Studies

The chosen case studies were action-oriented, social movement-based, and concerned with food and its production, consumption, and distribution in some ways. Case study activist-researchers, working in the groups associated with the case study, were invited to research and write the individual case studies and to participate in the collective analysis workshop at the end of the process.

Navdanya: A Grass Roots Movement in India to Conserve Bio-Diversity and Sustain Food Security

The Navdanya (nine seeds) project was based in three locations in India: the Garhwal Himalaya, the Karnataka Deccan, and the Western Ghats. Three local social movements were key to raising awareness and carrying out the work of Navdanya: Chipko (Garhwal Himalaya); Karnataka Rajya Ryota Sangha (Deccan); and Appiko (Western Ghats). The project goals were to create conservation centres in each of the three areas which would in turn document existing bio-diversity, collect crops for storage, select seed varieties based on yield and resilience, propagation of seed varieties, demonstrate lessons learned, and facilitate building a farmer-based seed supply. Vanaja Ramprasad was the researcher responsible for this case study.

El Daen—Environmental Conservation in Western Sudan

The El Daen project located in southern Darfur, western Sudan, was an example of a locally initiated and community-based movement to both protect nature and facilitate appropriate land and water use in a wide variety of activities. It highlighted the role of women within their community in adapting to harsh environmental conditions within a framework of subsistence sustainability. The participatory learning process created a 'code of behaviour' towards all living things. This code stipulates activities which are prohibited, such as the cutting of large green trees, illegal hunting, use of traps, cutting trees near the water, and so forth. Simultaneous work on local scientific experiments to develop ways to store seed cake indefinitely at low cost was also successful. The late Dr. Naila Babiker Hijazi was the researcher responsible for this study.

Berlin and Brandenburg as Centres of Environmental Activism: Organic Food Consumption and Organic Gardening and Farming

In Germany, popular or non-formal environmental education efforts have brought about a lifestyle transformation among many Berliners. 'Bio-shops' (organic health food shops) have become popular around the city. Vegetarianism and eating locally grown foods from organic cooperatives is becoming more and more common. Citizens' groups, most often spearheaded by women, have lobbied governments and raised public awareness to lower speed limits within the city limits and to install bike lanes and block certain city streets to automobile traffic. Elizabeth Meyer-Renschhausen has been a member of this movement and was responsible for writing up the experience.

Food, Aboriginal Ownership, Empowerment, and Cultural Recovery at the Six Nations Community in Canada

At the Six Nations Aboriginal community in Canada, a local economic development and cultural recovery project has brought together Aboriginal immigrants to Canada with First Nations persons from the Six Nations Community in southern Ontario in central Canada. Their collective work has investigated traditional food crops, reintroduced many such crops to the community, and educated community members and young people about heritage foods, and to produce foods for sale to the community and beyond. José Zarate, the project co-coordinator originally from Peru's high plateau, was the researcher responsible for this case study.

Women's Citizenship in Action: The Struggle Against Hunger and Poverty and in Defence of Life in Brazil

Spearheaded by a women's network called *Rede Mulher*, women from slum areas of Brazil took leadership in a broad campaign to plant organic gardens near where they live, to take action on water and waste issues in their communities, and to analyze the links between gender oppression and environmental destruction in their communities. Links to global issues and national movements were made explicit through popular education practices. Moema Viezzer and Teresa Moreira were both involved in the women's environmental movement and were the researchers for this study.

Transformative Learning in the Venezuelan Urban Amazon

The growing urbanization of the Amazon is a result of the physical and economic displacement of indigenous peoples throughout the Amazon basin. As capitalist resource extraction and agri-industrial practices proliferate, millions of people have been uprooted and forced to move, to live impoverished lives in the new cities of Brazil and Venezuela. This case study looked at the learning dimensions of a local movement in southern Venezuela to recuperate indigenous uses of the Amazon palm as a source of food, housing material, and medicine within an urban context. Traditional indigenous knowledge is seen as a contribution to a better quality of life, as well as to a recovery of pride in the traditional origins of many of the urban immigrants. Omar Vallez was both a leader of the original project and the case study writer.

People's Rights, Environmental Education, and Ecological Action for Sustainability in El Salvador

The growth of social movements in El Salvador is part of the recovery process of a country that had undergone a brutal 12-year civil war where environmental degradation was a strategic part of the suppression of peasant (often indigenous) farmers. The work in this project involved rebuilding local ecosystems through popular education with local materials and in local languages. It represented the collective action of women's groups, social researchers, popular educators, and the indigenous social and political leadership to begin the process of recovery as a people within a larger ecosystem. Marta Benavides was both an organizer and researcher for this project.

Collective Analysis Workshop

The data generated by a series of working papers, the case studies, and the first stages of an international survey were discussed at a collective analysis workshop which brought together the case study writers, the research coordinating team, and additional research assistants. The collective analysis workshop, itself a powerful example of transformative learning, placed the experiences of the environmental activists-researchers at the centre of the analysis and, through a variety of participatory research and popular education approaches, reflected on the role of learning in social movements.

What Was Learned?

A single chapter cannot do justice to the richness of the interactions, the depth of experience shared in the project or the full sense of transformative vision which passed among the project team at several moments during collective analysis workshop. However, it is possible to provide a summary of four key themes that emerged: principles of transformative environmental adult education, practices, and indicators of success.

As Hall and Sullivan (1993) noted in their conceptual paper for the study, transformative learning contains moments of critique, survival, and creation. It is about working with a diverse set of concepts and following principles in the intentional creation of educational practices and processes. But how does learning interact with social movement contexts? What roles do the various educational approaches embedded in the day-to-day working of social movements play in strengthening processes of action? After examining the various case study experiences, the collective analysis workshop arrived at a number of key principles which emerged from the practices themselves. These principles are a useful framework for thinking about all forms of transformative environmental adult education.

Recovery of a Sense of Place

The propensity to destroy the ecological balance in our communities varies, in part, according to the degree of 'sense of place' which we have. Place refers to our locations in bio-regional terms and also in terms of such social indicators as race, class, gender, sexual-orientation, and able-bodiedness. As Meyer-Renschhausen (1992, 8) noted in her case study in the *Awakening Sleepy Knowledge* report of linking organic vegetable growers to Berlin consumers, the members (of the food co-op) now know exactly where their cabbage comes from. As a result, we concluded that principles of bio-regionalism are important to developing a sense of place, and that we need to think of ways of building practical and theoretical ways to recover our sense of place when planning learning experiences.

The Importance of Bio-Diversity

Bio-diversity is that complex celebration of difference that allows for the flowering and survival of the world. Respect for bio-diversity means honouring space for bio-diversity to flourish. Bio-diversity performs its magic best when performing in settings that most humans understand as wild. As Vanaja Ramprasad (1992, 9) noted in the case study of Navdanya, bio-diversity is vulnerable, and left unprotected it tends to erode. The reduction of bio-diversity in the form of fewer seed varieties, extinction of animal species or the disappearance of other life forms threatens our survival. The full implications of concepts such as bio-diversity have broad meanings even for our understanding of the roles of our particular human species. Respect for education of a transformative variety increases the visibility and understanding of the importance of bio-diversity in ways that make sense in the particular context involved. Again, from Ramprasad, a conservation of bio-diversity and crop varieties *in-situ* on farmers' fields is a security imperative in the context of the North-South conflict over genetic resources (13).

Reconnecting with the Rest of Nature

Our pedagogical practices, according to our understanding of transformative learning, need to seek specific ways for us to reconnect with the rest of nature. As Viezzer and Moreira (1992, 17) say in their case study of the Jardim Kaghora community in Brazil, it is necessary to share the joy of living without domination among human beings or between human kind and nature. The first aspect of this is to recognize that we are part of nature and not apart from nature. We are connected with every form of life as we share the same molecular building blocks. Our ability to survive as a collectivity of all living beings depends on each of our species surviving in ecologically interconnected webs of life. This means that opportunities of a theoretical, practical, experiential, and participatory nature need be sought so that everyone can begin to recover a sense of the natural.

Awakening 'Sleepy Knowledge'

Increasing attention is being paid to the role of indigenous knowledge, even within academic settings (Dei et al. 2002). The concept of 'sleepy knowledge' came from the Venezuelan Puerto Ayacucho movement for the recovery of traditional environmental knowledge to help urban indigenous migrants cope better with the new conditions facing them. As knowledges and system of thinking have come to be so dominated by Eurocentric, rich country, patriarchal paradigms, older and non-dominant forms of knowledge have been allowed to 'go to sleep.' Ovalles (1994) describes the educational process of 'awakening' being done in Puerto Ayacucho as:

> a social process through which the values, principles, knowledge, etc. learned from the practices of past generations and found in the personal and collective consciousness of people are critical. These values, principles and knowledge come from the experience and relationships between societies and their natural environments throughout history. Due to the socialization process, this knowledge has been lost, and no longer transmitted from generation to generation until now. (2)

In addition to the knowledges of ancient peoples, the knowledges of women and of those who live closest to subsistence have much to offer us for environmental adult education. As the keepers of seeds, primary caregivers in communities, farmers, haulers of water and wood, and vibrant social and environmental activists and educators worldwide, many women bring more life-centred visions and ideas to environmental discourse.

Acting and Resisting

Facilitating action and supporting resistance are key principles for transformative adult environmental learning (Clover 2002). As Ovalles (1994, 4) says about the work in the urban Amazon, "learning becomes transformative in the moment that it starts to influence power, work, management and cultural relations." It might well be argued, for example, that even western science, with its built-in biases, offers us sufficient proof of the declining health of our biosphere. But that knowledge alone cannot help us if it is not linked to social and political actions that can make changes in the laws or practices which destroy us. Resistance, itself a form of action, is that quality which allows us, as individuals and as collectivities, to maintain our sense of integrity and community thereby denying others of power over us in important ways (Clover 2002). Transformative learning seeks out action and supports resistance.

Building Alliances and Relationships

In each of the examples of transformative learning that we researched, there was a strong emphasis on the importance of people working together. This is because change of a systemic nature is a long-term matter that requires skills and energies beyond any single person. Each of our cases of transformative learning involved the creation of alliances across diverse groups. In Sudan, the rural environmental association created an alliance with adult educators at the University of Khartoum. In Brazil, popular organizations of street kids, workers, women, and others came together in a poverty and hunger campaign. In El Salvador, former members of the armed opposition established new alliances with peasant leaders. Transformative learning needs to find ways to strengthen our skills in working with others. It has to do with organizing, understanding difference, respecting diversity, learning how to build consensus, reaching out to those who do not share our views, and with sustaining long-term political and operational strategies. This may be shantytown women coming together to start a food bank. It may involve *campesinos* in El Salvador eating together or joining the Rural Leadership Network. Ovalles (1994, 4) says that, in Puerto Ayachucho, "at each meeting they tried to make up networks of individuals and organizations which would permit continuing of the process"; while in Brazil, Moema Viezzer and Teresa Moreira (1992, 9) noted that "In November last year we organized a committee which has worked on three fronts all along. We established a bridge between middle-class schools and committees from middle-class apartment buildings and committees against hunger set up by the Neighbourhood Association from Jardim Kahohara (the slum community)." For us as environmental adult educators the question is: How can we put into practice ways of learning which strengthen the building of relationships?

Skills are Important Too

Transformative learning is not just about understanding concepts and connections; it is also about learning and teaching specific skills. Words such as *empowerment* sometimes obscure the fact that specific skills are involved in environmental action, and that learning how to do something may be as empowering as a new insight that gives broader meaning to one's daily life. In the Six Nations of the Grand River in Canada, for example, learning how to farm in the traditional way of the ancestors involves skills as well as consciousness. In the Sudan, "Women started to exchange information in ways and means of preserving food" (Hijazi 1992, 10). Successful organic bio-shops in Germany require skills in running a small business. Similarly, several skills were needed in the Navdanya project, including "cleaning and documentation of seeds, seed conservation, varietal improvement, in field agronomy trials" (Ramprasad 1992, 16) The challenge to those of us who work or seek to work with transformative forms of learning in these contexts is to identify the specific skills needed for the actions intended and to arrange ways to learn them.

Valuing Process in Learning

Many of our most unsuccessful educational experiences have focussed on trying to get the most 'facts' across in the shortest period of time. In transformative learning, however, the process of the learning is as important as the content of learning. Beginning with the daily lived experiences of those involved, transformative learning calls for increased attention to the relationship of the learning processes to the overall goal of our work. As Ramprasad (1992, 18) noted:

In the Navdanya project this is referred to as evocative forms of training. Instead of trainers transferring knowledge and information as if into an empty cup, the trainer draws out the wisdom that is lying dormant within the vast range of agricultural experience that the farmer has. By doing this the active thinking is awakened within the inner life of the farmers.

Deconstructing Relations of Power

Learning for transformative purposes involves understanding relations of power within a specific context. Understanding relations of power helps in understanding the exploitation or abuse of nature and people in particular situations. For example, an important part of village level seed projects in the Navdanya project in India involved understanding the relations of power within the Indian state, agri-business global corporations, and the sale of hybrid seeds and fertilizers. A seed conservation project would not be effective if the conservation practices were learned in the absence of any understanding of such relations of power. Benavides (1992, 3) noted that, in El Salvador, many farmers had to use their small plots as collateral for loans to buy seeds or fertilizer, and soon found that they lost their land as soon as they were unable to make repayments. Understanding the relations of power also allows for the potential to alter those relations and, most importantly, provides a framework for analyzing future actions by agri-business interests. Power flows through each and every practice in everyday life via gender relations, race and ethnic relations, class relations, and more. Sensitivity to the complex relationships of power and knowledge in ecological contexts is a goal for transformative learning.

The Practices and Processes of Transformative Learning

The collective analysis workshop process generated nearly one hundred specific practices and processes of transformative environmental adult education that had been used within the case studies. Many of the practices that formed the heart of the environmental actions under discussion had not been intentionally designed as educational practices, but became key moments for very powerful learning which deepened the understanding of the actions at hand, and reinforced the sustainability of the overall work. Many of the principles identified in the preceding section are incorporated in these practices. Two or three examples from each of these categories are used to illustrate the diversity and creativity of the practices. The practices identified include: celebrations and rituals, 'on-the-spot' learning, learning from elders, community meetings, nature tours or study visits, gender analysis, medicinal plant collections, kitchen composting, marches and protests, and the creation of community markets.

Celebrations and Rituals

Celebrations and rituals represent an important form of environmental adult education. They have the capacity to combine new and old knowledges, spiritual and physical activities, and various ways of coming closer to the Earth. Meditation was also used in some of the case study contexts. In making use of celebrations and rituals for environmental education, we are drawing on some of the most powerful and ancient methods we know. The following are but two of literally thousands of such activities.

The Navdanya project in India saw transformative learning as an awakening of the spiritual facilities that slumber within the farmer. Learning is a drawing-out of the wisdom lying dormant within the vast range of agricultural experience of the farmer. One of the practices used by Navdanya was to integrate the work of identifying and preserving the best genetic stock for the coming seasons into traditional seasonal festivals. In this ceremony women played the central role as 'keepers of the seeds,' as they participated in an elaborate festival of song and dance that focussed the entire community on the process of seed preservation. The seeds that had been selected for keeping were identified by the local farmers who elaborated their own criteria for selecting seeds based on qualities derived from their very specific contexts, including the desire to grow without imported fertilizers.

The Rural Leaders Network in El Salvador believed that their cultural, spiritual, and agricultural work feeds empty stomachs, supports the community, and feeds the soul. Noting that the spiritual aspect of life must not be underestimated, they made extensive use of song and music in their training programs. They also pay particular attention to sharing traditional histories when they begin a training workshop by asking leaders from the different communities to tell the stories of their communities. These are communities with many thousands of years of history, and the telling and exchanging of each story calls forth the years of resistance of the people of the region, and shows common elements among them.

A number of small celebrations were also held throughout the collective analysis workshop. These culminated at the closing ceremony through collectively weaving a web of sisal twine, pausing each time to think about what our experience had meant to us.

'On-the-Spot' Learning

A second broad area of practice that emerged from the collective analysis process of the project was labeled 'on-the-spot' learning. In India they spoke of *in situ* learning, as opposed to *ex situ* learning which was compared to the practices of *in situ* preservation of seeds compared to *ex situ* preservation of seeds. It was pointed out that maintaining control of the genetic stock is being done more responsibly by the communities than by private companies.

If we think about educational work, the same thing might be said—education works best when it is kept close to the communities and suffers when it is designed at a distance by others. In all of the case studies we examined, a majority of the learning was done in the farms, homes, shops, workplaces, or elsewhere where work and daily life was going on. The links between action, relevance, and natural processes are so much more obvious when one is standing in the field.

An example of on-the-spot learning was identified in the Community Garden, an integral part of the project at Six Nations in Canada. When Leman Gibson, the Mohawk Elder, works with the trainees at Six Nations of the Grand River near Brantford in Southern Ontario in Canada, he does so on the gently rolling land behind his home, which is one of two community gardens and test sites for growing traditional varieties of indigenous foods. His stories grow out of the land much as the corn, the squash, and the beans (these three foods are known traditionally in Iroquois culture as the 'three sisters'). Whether thanking the Creator directly in a prayer before working, or simply through the respect He shows for the Earth which brings so much each year, the trainees get spiritual, technical, and philosophical learning while they work on the spot.

In Sao Paulo, Brazil, as part of the large-scale food for life campaigns of the mid-1990s, a practice of reclaiming urban space for gardens was established. Homes are small and close together in the shantytowns of urban Brazil and, while we may be used to thinking of gardens and farms as rural experiences, the women's groups in the Hunger Campaign re-appropriated urban space to create gardens for the cultivation of what they called 'seeds of diversity.' In spaces that have usually been used for garbage, women have reclaimed the land for gardens where they grow different crops and share what they grow. It is very difficult to provide security for crops in a community where hunger is prevalent, but the urban gardens provide more than just a source of food. They are at the same time a place for women to come together to figure out a variety of other ways to survive. The gardens also offer ideas about other more productive uses for the urban land that all too often becomes just another dump.

Learning from Elders

To some extent all of the case studies examined had an element of learning from elders. In Germany, the urban women and men of Berlin went to some of the older organic communes in the surrounding countryside to learn from the elderly farmers. In the Six Nations agricultural project two elders, one man and one woman, are recognized as key advisers for the project and as teachers on practical farming and gardening techniques. The members of the Six Nations community could not work at a community level without the involvement of some of the elders. Similarly in the El Daen area of Sudan, the Elders (all male in this case) were at the heart of the conservation decisions and leadership.

Who Are Our Elders in the Various Educational Contexts That We Work in?

Community Meetings

Community meetings of both an informational and participatory nature were common across all the case study projects. Transformative learning made full use of the very wide variety of popular education and adult education methods in these community settings. These included cinema study groups, storytelling, cartoons, posters, community theatre, role-playing, song, music, and art. The Brazilian campaign and the work in El Salvador, Venezuela, and India all made rich and imaginative use of participatory approaches for community meetings.

Nature Tours or Study Visits

Nature can teach us much if we can learn to hear its messages and read its signs. Sometimes other species and plant life are all around us. In these cases educators need to learn to be quiet and become more open to the world around. Sometimes in urban settings we feel distant from and apart from the other parts of our natural world. In these cases it is useful to go to places where nature is more healthy and sustainable from an ecological point of view.

One of the study visits in Sao Paulo involved taking the women from the shantytown to the area's water treatment plant. Here it was possible for the women to see all the chemicals that are put into the city water and to talk with the workers about how water becomes contaminated in the first place. The results of this visit were clearly seen when the women's group reclaimed a spring in their neighbourhood, removing all the surrounding rubbish and putting up a barrier around it so that it would be kept clean. In Berlin, urban environmental activists in search of a way to strengthen their relationship with nature began to make visits by bicycle to organic farms, in both the former west and former eastern parts of Germany around Berlin. Similarly, study visits of townsfolk to villages were organized in the Indian case study where one of the objectives is to educate the consumers of various food products.

Gender Analyses

Women and their lives are at the heart of transformative learning. Understanding the differential impact of environmental destruction on the lives of women is critical to being able to find a solution in most cases. In all of the case studies, women, because of their central role in food growing, food preparation, and care-giving in general, were fundamental to transformation. Practices such as consciousness-raising, feminist popular education, cooking classes, and other activities that simply allow women to be together outside of the very heavy load of daily work were found in most of the case studies. This means that educational practices that directly or indirectly allow for increased visibility of the roles of women, particularly among men, are important.

Medicinal Plant Collections

Plants were used for various medicinal purposes in the case study projects in El Salvador, Six Nations of the Grand River, Venezuela, Germany, and the Sudan. In El Salvador, in particular, the growing of medicinal plants is a central part of the rural leadership work. Recovery of knowledge about medicinal plants strengthens the self-sufficiency of communities and reduces the dependence on expensive imported medicines for common ailments. There has been a revival of such interest in each of the countries, and the active tending of such herb gardens has proven to be an excellent adult education tool.

Kitchen Compost Piles

In Brazil and Germany the keeping of a kitchen compost pile was an intentional part of the work of environmental activists. The use of a compost pile, with its direct and visible lesson about reducing organic waste and the power of nature to nourish itself by turning waste into good soil, is one of the best ways of letting nature teach us. Along with reducing our creation of rubbish, learning to recycle organic products can make a substantial difference in our communities.

Community Markets

In both the Berlin and Six Nations case studies, creating a community market was seen as an educational as well as an economic activity. Green activists in Berlin have been prevented from using plants for medicinal purposes or from buying local organic produce by the large food producers and food marketing organizations that have historically held a monopoly over food distribution. The creation of neighbourhood 'bio-shops,' where local producers can find buyers, has proven to be both a functional marketing arrangement as well as an excellent place for informal education to take place. The buyers, sellers, and shopkeepers educate themselves about many issues having to do with ways to live more lightly on the Earth. At Six Nations, market research at the start of the project indicated that the community spent nearly $11 million dollars a year for food. None of that money was spent in the community. There were plans for an integrated community market and educational centre as part of the recovery of traditional food crops, and for better use of the land in the community. This centre would provide the community with films and talks about aspects of indigenous agricultural and culture, as well as being a destination for visitors wanting to buy locally, and wishing to learn something about the area.

Marches and Protests

Environmental action takes many forms. Sometimes it is important to take collective action in order to bring certain information to the authorities involved. Marches, protests, and other forms of non-violent action are powerful ways to learn about the distribution of power, the role of different kinds of knowledge, and the strength that comes from acting together. Whether in India protesting the patenting of the Neem tree, or in Brazil going to local authorities to provide clean water, the right to take collective action is one of the most powerful learning tools and a means of responding to unfair environmental practices available.

Indicators of Success: Social Movement Evaluation of Learning

The concept of 'indicators of success' was criticized as reflecting the language of top-down externally driven training or education by some of the participants at the workshop. But, as discussion continued, participants became more comfortable with a discussion about what to look for in social movement learning contexts, and agreed that, under certain circumstances, thinking through the question of longer-term goals and objectives for the learning process was valuable. It must be stressed that the specific contexts are the key to any question about success with the fundamental question being success for whom? The question of 'success' needs to begin with the women, children, and men whose lives are at the heart of the specific processes in question.

Nine 'indicators' or ways to know how well the environmental adult education work is proceeding were identified:

- the development of new practices;
- increased participation or mobilization;
- changes in gendered roles or behaviours;

- linking between local and global contexts;
- production or recovery of knowledge;
- new legislation or policies;
- increases in self-sufficiency and bio-regionalism;
- increases in cooperation; and
- existence of new alliances and networks.

To what extent have new practices surfaced or developed? This obviously differed in each context. In Berlin, the citizen groups involved with greening the city wanted to see more bicycles on new bicycle lanes. In Sao Paulo, Brazilian women wanted to reclaim land for urban gardens, so the number of new gardens is an important fact. In the Indian case, farmers refusing to buy fertilizer and hybrid seeds is a new practice indicating a new awareness. In Puerto Ayacucho, Venezuela, increased use of palm trees constitutes a new practice. Each environmental action will have potential to support new practices of very different types.

A second change worth noting relates to participation and mobilization. Mere participation is not, in and of itself, much of an indicator. However, increased participation in specific new practices or in activities based on local mobilization for change can be a very important indicator. In a large-scale campaign, such as the Poverty and Hunger Campaign in Brazil, increased participation was a critical ingredient in the overall effectiveness of the project. In village seed conservation projects, the number of families participating in the seed conservation program was a valuable objective of the project. One could, of course, turn this indicator around and say that decreased participation in agri-business programs, or farming practices dependent on artificial fertilizers, would be the measure of success. Participation needs also to look at the question of who is participating. What is the role of indigenous peoples or women in a particular process?

Changes in gendered roles and behaviour are a third aspect for qualitative evaluation. To what degree are women treated better than before? Do more men begin to stay in the community and help or do they continue to leave? Do boys and girls begin to share roles in more equal ways? Is there more awareness of the differential impact of toxins, pesticide poisoning, and unsafe food supplies on women and girls? Are women given support for the additional work they take on in providing energy for environmental action projects?

A fourth indicator of success involves connections being made between local and global contexts. To what extent are people making the connection between the seed that they hold in their hands and the global struggle over ownership of genetic stock? Can people understand the link between hunger in their homes in Sao Paulo, and the expansion of business interests into sugar cane production for alcohol for cars in the cities? Why are the varieties of food crop species declining? Why are traditional uses of the palm tree being lost due to artificial substitutes? What does twenty-five years of war mean for the ecosystem of El Salvador?

The degree to which 'sleepy knowledge' is being recovered, and new knowledge produced, is a fifth indicator. Our participants said that one indicator of success would be when academics begin to learn from local people. We might want to add—and when local people get credit for the knowledge! Are new uses of plants revealed or rediscovered? Are new understandings of relations of power which keep change in environmental matters from changing noted? Are new ways of relating to other animals and plants learned? Are new ways of gardening, composting, and preserving foods learned? Have new ways of understanding the role of non-violent action been identified? Have citizens produced new knowledge about the impact of consumerism on their own lives?

Another sixth way to judge what is happening is the degree to which new legislation and regulations that support changed environmental practices are established. Is local legislation introduced to control environmental degradation and is the legislation enforced through a court or similar bodies? Are international rules and normative statements about bio-diversity enacted locally? Are new codes of behaviour based upon sustainable relationships between people and other living things being developed? Are laws regarding levels of pesticides in our food monitored and enforced? Are people's rights to land enforced by law? Is the right to learn respected? Has the environmental action moved from practice to law?

A seventh indicator identified by the study dealt with increasing self-sufficiency and bio-regionalism. To what degree are local practices being found for local issues? To what degree are local farmers independent of global suppliers for agricultural inputs? To what degree do local farmers supply local populations with food? To what degree does international information available support community-based decision-making? To what degree are land use issues based on the specificities of the bio-region in question. To what degree are people de-linked from the global market?

The degree to which cooperation among people increases, in the interest of sustainable solutions to issues of food, is an eighth factor. Are there new structures for cooperation over marketing, seed sharing, using traditional products, solving as yet unidentified issues? Has people's isolation in the face of unjust environmental practices been reduced? Are individuals able to get support for struggles against landowners, unfair government regulations, or other forms of oppression?

The ninth factor to take into account may be one of the most powerful. To what degree have new alliances and networks been created, encouraged, or supported? To what degree are local communities able to gain solidarity, or share local knowledge with others in a broader alliance? To what degree are local activists able to make use of information available on national or international networks to strengthen their own actions?

Conclusion

We have a rich global heritage on the role of learning in social action. We have the lessons of Ghandi-ji during the Indian Independence campaigns, of Julius Nyerere of Tanzania of the 1960s, 1970s, and 1980s, who conceived of education playing the key role in the creation of a new self-reliant African nation, and of Paulo Freire of Brazil, who taught us that education is never neutral. In the late nineteenth century in Europe the newly emerging social movements of labour, women's suffrage, and peace developed intentional adult learning strategies within the context of their overall movements. More recently, we have the voice of Vandana Shiva raising concerns about global capital's attempts to patent life itself; and of Mathew Coon Come, the National Chief of Canada's Assembly of First Nations, whose Cree First Nation in northern Quebec won a major victory over the combined interests of the state and private capital planning to build a hydro-electric dam on Cree land. In history, as in our contemporary contexts, the learning dimension of social movement transformation has been of critical importance.

The Transformative Learning through Environmental Action Project built on both these historic and contemporary dimensions. The people involved in this project followed and expanded the pathways that others have travelled before. The study offers us, to my knowledge, one of the only pieces of research that is an intentional exploration of the learning dimension of six important environmental action campaigns. It brought a group of activist-intellectuals together from environmental social movement contexts in Brazil, El Salvador, India, Sudan, Canada, Germany, and Venezuela. It also offers data from the case studies and the collective analysis process to help all those concerned with the growing theory and practice of environmental adult education. It tells us social movements are, among other things, about creating pedagogical spaces for adults to learn to transform their lives and the structures around them. It tells us that attending to the learning dimensions of environmental activism is a critical component in social movement strategy. It tells us that gender, place, global power, and local creativity are inextricably linked in life and must be understood, worked with, challenged, and celebrated as part of environmental adult education. It also tells us that, far from the media and far from centres of global financial power, millions of women and men are engaged in learning and taking action each day to survive and create new relations. When the house of cards that is the global economic system tumbles down, as surely it must, the seeds of a new world are already planted.

References

Benavides, Marta. 1992. "People's Rights, Environmental Education and Ecological Action for Sustainability in El Salvador." Case Study in "Awakening Sleepy Knowledge: Final Report." Toronto: Transformative Learning Centre.

Clover, Darlene. (2002). "Traversing the Gap: Consientizacion, Educative-Activism in Environmental Adult Education." *Environmental Education Research 8*, 3, 315–323.

Dei, George, Hall, Budd, and Goldin-Rosenberg, Dorothy. 2002. Second Edition. *Indigenous Knowledges in Global Contexts: Multiple Readings of the World.* Toronto: University of Toronto Press.

Hall, Budd and Sullivan, Edmund. 1994. "Transformative Learning: Contexts and Practices." In "Awakening Sleepy Knowledge: Final Report" for the International Development Research Centre, Ottawa.

Hall, Budd L. 2000. "Global Civil Society: Theorizing a Changing World." *Convergence XXXIII*, 10–32.

Hall, Budd L. 2000. "The International Council for Adult Education: Global Civil Society Structure." *Convergence XXXIII*, 33–56.

Hijazi, Naila. 1992. "El Daen—Environmental Conservation in Western Sudan." Case Study in "Awakening Sleepy Knowledge: Final Report." Toronto: Transformative Learning Centre.

Holford, John. 1995. "Why Social Movements Matter: Adult Education Theory, Cognitive Praxis and the Creation of Knowledge." *Adult Education Quarterly 45*, 2, 95–111.

Meyer-Renschhausen, Elizabeth. 1992. "Berlin and Brandenburg as Centres of Environmental Activism: Organic Food Consumption and Organic Gardening and Farming." Case Study in "Awakening Sleepy Knowledge: Final Report." Toronto: Transformative Learning Centre.

Mojab, Shahrzad. 2000. "The Feminist Project in Cyberspace and Civil Society." *Convergence XXXIII*, 106–119.

Ontario Institute for Studies in Education. (1994). *Awakening Sleepy Knowledge*. Ottawa: International Development Research Centre.

Ramprasad, Vanaja. 1992. "Navdanya. A Grass Roots Movement in India to Conserve Biodiversity and Sustain Food Security." Case Study in "Awakening Sleepy Knowledge: Final Report." Toronto: Transformative Learning Centre.

Tandon, Rajesh. 2000. "Civil Society, Adult Learning and Action in India." *Convergence XXXIII*, 120–137.

Viezzer, M. and Moreira, T. 1992. "Women's Citizenship in Action: The Struggle Against Hunger and Poverty and in Defence of Life in Brazil." Case Study in "Awakening Sleepy Knowledge: Final Report." Toronto: Transformative Learning Centre.

Welton, Michael. 1993. "Social Revolution Learning: The New Social Movements as Learning Sites." *Adult Education Quarterly 43*, 3, 152–164.

Zarate, José. 1992. "Food, Aboriginal Ownership, Empowerment and Cultural Recovery at the Six Nations Community in Canada." Case Study in "Awakening Sleepy Knowledge: Final Report." Toronto: Transformative Learning Centre.

Chapter 12

The Nature of Transformation:
Developing a Learning Resource
for Environmental Adult Education

Darlene E. Clover, Canada
Shirley Follen, Canada

Human dialogues with the Earth and sky vary with the relationship humanity has to
its environment. (Kane 1998, 70)

It has been suggested that "much energy and money is spent in the
production of environmental education materials world-wide" (Camozzi in
ICAE/Ecologic 1994, 39). This means that, for working with children, there
exists a wealth of resource materials with new principles and creative
activities (for example, see Pike and Selby 1988; Cohen 1989). There is also
a vast and important collection of learning resources to assist educators,
practitioners, community animators, researchers, and activists working in
the field of adult education and learning. Emerging particularly from the
streams of feminist, popular, and labour education, there are diverse and
creative activity-based adult education texts which deal with issues from
gender and race oppression to 'humanizing' the workplace (for examples,
see Arnold et al. 1991, Bishop 1988; Nadeau 1996).

However, when in 1995 we took on the challenge of working with
diverse groups and communities across Canada and worldwide, and weaving
environmental issues into the fabric of our adult education theory and
practice, we discovered that few resource materials existed in either
environmental or adult education which could assist us. Therefore,
developing a new dimension of ecological adult learning posed a very real
challenge to our abilities and creativity as adult educators. As adult
educators often do, we decided to develop resource materials in the form of
a book, but in one of the most complex and challenging ways: Collectively,

through a participatory process which extended from one end of Canada and beyond to other countries of the world.

This chapter is an examination of some of the elements involved in developing that learning resource. *The Nature of Transformation: Environmental Adult Education* (Clover et al. 1998 and 2000) brings together social learning theories with environmental thought and issues in order to broaden the framework, theory, and practice of adult education. We begin with the context that includes the need for this type of book, a brief description of some of the places or spaces within which we worked, the groups we worked with, and how we worked with them. We also weave together some of the educational principles and theories that guided our practice with contemporary ideas of how and where environmental adult education resource materials should be developed, linking this to our process. We then explore how and why various activities emerged, and include an analysis of some of the successes of this work but, also, where and how we failed from time to time. The chapter concludes with a few of the lessons we learned in developing this book which we hope will be beneficial to those who decide to engage in this challenging form of resource creation.

The Context

The Need: Merging Environmental and Adult Education

In 1986, UNESCO argued that:

> [I]t has been accepted that the educational system like the rest of society has not been able to adapt quickly enough to rapidly changing environmental conditions. This situation is unlikely to change dramatically in the next decade and the world cannot wait for the new generation of politicians and decision-makers to emerge. It is necessary to develop an effective nonformal [adult] education process. [This] is of the utmost importance in creating a society whose citizens are knowledgeable enough to make a valid contribution to decision-making processes. (8)

Adult education has broadened over the past few decades to encompass issues such as race, welfare, employment, and gender (Thomas 1991). It has also demonstrated an ability to engage within dynamic discourses such as globalization, literacy, and global activism. However, it has been somewhat negligent in grasping the significance and complexity of environment issues in terms of their links to social, political, emotional, spiritual, and cultural dimensions.

Much of theory and practice of 'environmental education' has been designed for students and schools. When some believe they can extend these to the 'education of the citizenry' or 'public education,' it often becomes a matter of 'awareness raising' and 'individual behaviourial change.' The practice of 'awareness raising' most often translates into sharing more and more data or information, developing more and more information pieces rather than creating critical, active, and engaged processes of learning for adults.[2] The discourse of behaviourial change de-politicizes and privatizes environmental issues, focussing not on emancipatory knowledge creation and challenging relations of power, but rather changing personal behaviour and attitudes.

But in a world where many humans are losing their instinctive links to the Earth, and the implications of capitalist, patriarchal, cultural homogenist, and economic development ideologies on the cycles of life are so profound; there must be a weaving of these two fields. The fundamental Earth- and life-centred qualities of new forms of environmental education must be woven with the critical and social change-oriented aspects of adult education if we are to fully tackle the politics of environmental degradation with and through the adult population, or those with the most power in society, and bring about socio-environmental transformation. Therefore, we see the development of environmental adult education and this book as a political responsibility. We are not interested in behaviour change but in bringing people together around the politics of socio-environmental problems, to challenge, critique, laugh, struggle, and create.

We feel that this is the best way to complement and support the work of teachers and researchers attempting to radically re-shape teaching and learning practices within the classroom.

The Lack of Resources for Adults

Camozzi (in ICAE/Ecologic 1994, 39) notes that "absolutely the most important step in the production of any environmental [adult] education materials is to conduct some form of needs assessment research." It is also important that the adult educator engage in "a thorough search to find out if there are already existing materials that can be used or adapted."

It did not take much time, after defining ourselves as 'environmental adult educators,' a totally unknown concept in adult education, and setting out to integrate new ideas into the ecologically threadbare fabric of our field, to discover that there were few adult and/or environmental education materials which reflected exactly what it was we were searching for.[3] One reason for this is, environmental issues have not been viewed as either relevant or important to the field of adult education and have, therefore, never been a part of the discourse or practice. Until quite recently, like many other fields, adult education and its major theoreticians and practitioners have tended to be overtly anthropocentric (Bowers 1991; Clover 1999). A second reason is that the vast majority of environmental education materials have been written to work with children and "materials developed for children are not necessarily effective in the teaching of adults" (Burch 1993, 28). They are not effective because they are not designed to work with people who have the vast knowledge and experiences of adults. In addition, and for obvious reasons, they are more simplistic and, therefore, less challenging.[4]

In addition to the fact that the activities themselves were not appropriate, the language used in environmental education texts is primarily that of formal education. It speaks of and to a world of schools, pupils, lessons, teachers, curricula, and classrooms—the world of formal education. The language of adult education is often quite different. This is a world of participants and learners, educators, agendas, co-learning, facilitation, and diverse spaces or sites of learning. We raise this issue because language is extremely important. It can be inclusive but it can also be exclusive. It can empower, but it can also diminish or dis-empower. Language can bring people and ideas together or it can create distance. An excellent example of this is 'gender insensitive' language which works quite effectively in marginalizing and dis-empowering women.

In addition, the idea of 'school' or formal education for many adults is unsatisfying and even repellent. Some, particularly First Nations peoples, have had bad experiences with 'schools.' Others simply find it to be something to which they have no need or desire to return to. Most adults take part in continuing education or community-based learning activities to make personal and collective changes in their lives and communities, and engage in active discussions around issues of their choosing, not to be 'taught' or lectured to on pre-designed 'lessons.'

Methods of Resource Development

Leal Filho and Villa Bandeira (1995, 56) note, that in terms of identifying resources and materials for working with adults, one needs to consider certain implications. The first is 'information for whom' and 'their information requirement' because "although an editorial article on a certain environmental matter may not be suitable for a schoolchild, it may be appropriate for an adult." The second is 'information for what' that "relates to the aims of providing environmental information" (IBID: 56). Leal Filho and Villa Bandeira (1995) and Camass (in ICAE/Ecologic 1994) stress the importance of ensuring that local examples are used in the materials and/or that the content is directly linked to people's own experiences as a starting point, rather than about something more removed and then working towards larger global problems and their links to the local. In brief, materials based in place have the most meaning and thereby often better engaging the attention of the adult.

It is reasonably argued that "[a]dapting materials is much less expensive than creating new ones" (ICAE/Ecologic 1994, 39). But given that there are few materials for working with adults, environmental adult educators worldwide encourage that those which are developed for the adult education process be directed towards adults in the onset, "rather than adapted from those written for children" (Ibikunle-Johnson and Rugumayo 1987, 5). Environmental adult educators should, whenever possible, use indigenous or homegrown materials but also create spaces for the participants or learners themselves to collectively develop their own materials to better reflect community and/or regional problems and issues.

Camozzi (in ICAE/Ecologic 1994, 39) suggests that the "team production of materials" brings together different levels of expertise and gives those involved "a stake in the final outcome of the material." Producing materials collectively also adds a more creative dimension. Also, adults themselves should be made aware that they are creating knowledge and a new area of adult education, which will help others worldwide. Not only will the materials then "draw upon the vocabulary [language] universe of the learners" (Ibikunle-Johnson and Rugumayo 1987, 10) but those involved will feel more "empowered by their own knowledge and ideas" (Howell 1989, 52).

There are two other important elements for resource materials, particularly those that are practice-based. The first is that the activities should be accessible and easy to use, yet challenging. "It should be up to the individual instructors to modify and adapt the activities to their own needs and situations...not a prescription for how to run an activity...but a source of ideas" (Wright 2000, 42). This means that activities must be adaptable for a diversity of issues, contexts, and concerns. Second, the activities in the resource materials should always have "an element of 'fun' [for] adults like fun!" (ICAE/Ecologic 1994, 39). Like other issues such as poverty and globalization, environmental issues are broad, include dimensions of diversity and oppression, and are challenging and controversial which makes them difficult to deal with. People often feel confused by complex scientific debate, shrouded in jargon and mystery, such as discourses on genetically modified foods and cloning. They are also angry but confused by often 'corporate' messages that pit growth and employment against the environment and health as the vast majority realize that it is technology and the pursuit of profits that eliminate jobs and cause the most environmental damage. People are often frustrated around how to deal with the immensity of the problem and how slow things are to change, or just fatigued and apathetic with the entire problem. Humour, the ability to laugh at one's self and with each other, the ability to take 'time out,' can be a very empowering and fulfilling way to deal with these emotions.

The Nature of Transformation: Environmental Adult Education

What the Book Looks Like

> Human beings are thinking, feeling, creative and active beings that have the ability to acquire knowledge through various processes. And in every process there exists the opportunity to learn. The learning process is one of many integrated into people's lives. (Keough 1996, 3)

What we required was a book that could bring together all the activities we were collectively developing across Canada and around the world and, therefore, *The Nature of Transformation* became a predominately 'practice-based' text. It is about the 'process' of learning and education. Although it includes a focus on the learning and teaching principles, practices and theories of adult, popular, feminist, indigenous, and environmental education, most of the book reflects the practice of environmental adult education and ideas to develop workshops and other learning processes within this framework. The numerous activities are ways to draw out people's existing ecological knowledge and understandings of specific issues, challenge people's own assumptions and those of others, de-construct media portrayals of socio-environmental issues and dominant worldviews, and examine gender, race, class, and other relations of power in communities and the larger world within a framework of sustainability and transformation. They also engage people in critical discussion around globalization, corporatization and cultural imperialism, provide the opportunity to envision sustainability, reconceptualize work in a rapidly changing world, and develop short-, medium-, and long-term plans for action. The activities also use the community and the rest of nature as teachers and spaces within which to reflect upon issues and past experiences and relationships to and with the rest of nature in order to develop new partnerships in socio-environmental transformation.

The activities tap into the cognitive, but also the emotional, cultural, and creative dimensions of people. Many are geared towards celebration and empowerment. The activities in *The Nature of Transformation* are based in and on the issues raised by the communities with whom we worked. They are flexible and can be adapted to deal with any issue: from gender and race inequity to climate change, waste management to green fatigue, violence against youth to soil erosion, consumerism to human and ecosystem health. Although not impossible, it would be difficult and even counter-productive to follow them step-by-step since no two situations would ever be the same.

The Idea: Who We Worked With

> The most interesting discovery we made as environmental adult educators was that what ails people is not necessarily a lack of knowledge or awareness of environmental problems. In fact, most often people are highly knowledgeable... This fact became our most exciting challenge. (Clover et al. 1998 and 2000, 2)

The process of collectively developing this environmental adult education resource book began in 1993 when we were approached by a member of the Education Sub-Committee of the Ontario Healthy Communities Coalition (OHCC). This is a province-wide umbrella organization, which supports processes of community dialogue and collective action as the way to move towards a more healthy and sustainable world, and since the work is mainly with adults, they invited us to help them develop their educational strategy. We designed and facilitated four workshops in various parts of the province of Ontario, working closely in each area with the community animators. As we began to engage in this work, we realized that it should not be confined to Ontario and approached, or most often were approached by, a number of other agencies and organizations across Canada. Global dissemination of our work through the Learning for Environmental Action Programme, a global environmental adult education network, soon had us invited to work with others such as the Harmony Foundation in Victoria, British Columbia; the Canadian Foundation of University Women; the National Roundtable for Economy; the Environmental Education Association of Australia; the World Wide Fund for Nature in Asia/South Pacific; the Organization of Economic Cooperation in Paris; and various universities in Australia, Alaska, Canada, and England. Engaging with these differing and often contradictory groups made us better understand the diverse ways in which people defined the environment, and how much our educational methods and principles had to be re-shaped to complement or challenge this, as well as how much we actually required some sort of resource which we could use and share with others. Until we developed this book, we were continually printing off past agendas as guidelines, and telling those who were looking for actual books on this new area, that nothing actually existed.

On the Road: How We Worked

> Within any learning opportunity there is both content and process. The content is the topic, subject or issue. The process is the method used to examine the issues or topics. These two dimensions of learning are deliberate: together organizers and facilitators should identify and assemble the content and create the process. (Clover et al. 1998 and 2000, 2)

As mentioned, we were invited by a university department, community group, non-governmental organization, school, foundation, or someone who had met us previously at a workshop or somehow had learned about our work, to organize a workshop or course on the theme of environmental adult education.[5] Although most of the time these were inter-woven, there were often two distinct foci or goals to the workshops or courses. For simplicity's sake, we will refer to one as the 'issue-based' workshop or course. These provided the opportunity for participants to identify and critically examine the root causes of environmental problems they themselves identified in their community. The second we will refer to as 'educational-based.' The primary focus of this work was to learn through engagement about critical and creative adult educational approaches or tools, and in particular environmental adult education, in order to strengthen the educational dimension and capacity of those who attended the course or workshop.

In order to ensure that each workshop or course was based primarily on the context, needs, and objectives of the organizers and/or participants, we went through at least two drafts of each course/workshop outline, adding their suggestions and making changes. In addition to this process, we began each workshop or course by outlining the activities so that all those present could take ownership, understand the process, the purpose of each activity, and how the activities fit together. In addition, it was always understood as part of each workshop that we were working together to develop the theory and practice of environmental adult education, which would be put into a book to help others work within a contemporary socio-environmental politicized framework.

The Nature of Transformation was developed through engagement, discussion, debate, reflection and evaluation.

Where the Ideas for the Activities Came From

As mentioned previously, we had few resources regarding material and activities that focussed on the environment through a political lens and was suitable for adults. In the beginning, we adapted activities for children and 'greened' adult education activities.[6] But soon we began to create the activities ourselves based on a specific need, an issue, or an impulse. The latter, we would like to believe, derives from an innate creative instinct we share with many other adult educators and facilitators, and is by far the most gratifying when successful. However, as shown with the examples below, there have been exceptional successes as well as some activities that we either promptly changed or dropped for obvious reasons.

The basis for all our creations has been our long and diverse experiences working with adults, our love of the rest of nature, the belief that nature and community can teach, a realization that all issues are connected in some way or another, and finally, perhaps a small touch of the gambler in us. Ideas were born by using our emotions which flared when we saw a small shop give way to a Wal-Mart; when we heard the word 'mankind' used in the so-called inclusive way, when we read another report on the potential harmful effects of genetically modified foods, or when community members made links between complex issues we ourselves did not at first see. This latter meant that we relied heavily on the creativity and knowledge of the workshop participants, to reshape, redesign, and re-create the ideas.

Challenges We Faced

Each workshop we facilitated provided many challenges and new experiences. As adult educators, we were forced to find ways to introduce environmental learning strategies that would be acceptable and useful for adults. It soon became apparent that we were first and foremost adult educators and only one of us had taken any environmentally focussed courses or done a certain amount of reading in the area, and was therefore more familiar with some of the ideas of outdoor and environmental education. When, during the first OHCC workshop, a member suggested that she was going to ask people to actually go outside and engage in an activity she had been thinking about, one of the other facilitators flatly stated that "this will not work—adults don't want to go outside!" Since we had no experience with actually 'taking' adults outside, and it was a fledgling idea at best since it was simply not done in most schools of adult education, it was extremely unclear of what we should do. Therefore, the 'outdoor' activity became simply an 'add-on.' We suggested to the participants that when they went for a tea break, they should go outside and use their 'senses' to explore the natural world around them. Needless to say, no one did it! Although our workshops were about environmental adult education and the rest of nature, the latter was nowhere to be found and the former looked pretty deficient. All in all, we literally, as Paulo Freire did, collectively made the environmental adult education by walking.

We also discovered the importance of suitable workshop venues: rooms without windows, or access to outdoor space to take participants are definite inhibitors to a successful workshop that is designed to involve the rest of nature. One workshop was in a motel basement with no windows, not a supportive venue for an environmental adult education workshop. To remedy the problem, we drove out into the countryside and collected wildflowers which we placed in a basket at the front of the room. We then asked participants to draw images of the rest of nature with markers on flip-chart paper to make the room a warmer, and more vibrant and colourful space in which to learn and work.

We also learned about the dangers of having pre-conceived ideas about the participants. In one instance we were prepared for a room full of 'sophisticated' (Shirley Follen referred to them as 'pearls and heels') literary society women, whom we assumed would have little desire to step into an environmental adult educational process that might require getting their hands dirty or discuss relationships to the rest of nature. We were very wrong. The stories, the shared learning that took place in three short hours are forever reminders to us that environmental adult educators must be prepared for surprises. This reinforced the importance of knowing beforehand the needs and issues of participants.

A third challenge was trying to work together as a team, figure out each other's strengths. We each brought to our workshops varied, complementary backgrounds and experiences in the areas of feminist, popular, adult, and environmental education. Bringing these together naturally, working as one, took time and practice. It was important that we listened carefully to what each of us contributed, not only for the obvious reason that everything we do and say connects, but in order to evaluate and change where necessary. We accepted criticism as well as praise from each other, and learned how to meld our strengths. We also realized the benefits and importance or working with other facilitators, which broadens the scope of experience and knowledge, as well as the creative juices that are so necessary for successful workshop facilitators.

These Worked Well

Working in Harmony

We were invited to participate in the Harmony Foundation 1996 Summer Institute and presented two workshop sessions. The summer programme was held at the Lester B. Pearson College near Victoria, B.C., a beautiful setting by the ocean, overlooked by giant Douglas Firs. Approximately thirty-six people attended, with three facilitators and five session leaders. Ten countries were represented: Brazil, China, Guatemala, Ghana, Israel, Kenya, Mexico, Philippines, South Africa, and Canada.

The Summer Institute was designed to explore the connections between values, education, and behaviour change while recognizing the interdependence among social, economic, political, and environmental issues. Community, school, and workplace educators and leaders from around the world shared their expertise and applied new knowledge and skills gained to create innovative educational strategies.

The two sessions for which we were responsible were *Tools for Change* and *Educating for Change*. In the *Tools for Change* section we presented a workshop on critical thinking and using creativity. Participants examined a number of flyers from superstores in Canada and engaged in a discussion around values, waste, advertising, language, corporatization, cultural homogenisation, gender socialization, and so on. The session titled *Educating for Change* was meant to include all participants and serve as a bridge between the new learnings/skills sessions of the first three days and the two-day *Action Teamwork* session that was to follow. However, midweek we discovered changes had been made to the programme and we were forced to completely redesign our *Educating for Change* workshop. For the first three days, participants had been kept in specific groups. They had worked together, laughed together, challenged each other, and felt comfortable as a group. However, we discovered that the groups were about to be re-organized for the next two days and one of the other facilitators was planning, as a 'wind-up,' something tantamount to a wake! The problems that might ensue, not to mention the sorrows and sadness of this act of splitting the groups, was quite obvious. In addition, all of the facilitators and organizers were extremely nervous. We had to make it work.

Following our initial panic of 'what on earth are we going to do about this,' and 'how can we make this transition easier on everyone including ourselves,' we devised a workshop that not only met the new needs, but also made us realize that crisis and creativity go hand in hand.

We began by engaging the full group in an activity that would make them feel comfortable, relaxed, and allow them to have fun. We asked participants to make the sounds of an animal or bird or something else familiar from their country. We as facilitators took the lead with our own rendition of a Canadian bear growl that was followed by a variety of fascinating sounds from the group, including a strident first line from the Brazilian national anthem.

River of Learning and Transformation

Prior to the workshop we had placed on one wall a large piece of butcher paper on which we had drawn a river and backdrop. We moved everyone into their original groups, and asked them to engage in a process of reconnecting/synthesizing, to identify, share, and discuss the variety of educational philosophies and practices they had collectively experienced over the past few days. They were given rocks cut from brown paper, reeds cut from green paper, and so on, on which they wrote their ideas and then pasted them onto the mural. Together, they created a colourful visual of their time together.

After a short break, we allocated the participants into their new groups. This is where we decided that creativity and humour were required. We assigned each group a creative task. One group was asked to write a song called "Singing in Harmony," others were to prepare a skit, write a poem, and develop a short story based on one of the main issues that had arisen in the first few days. Reactions to using this creative approach were overwhelming. Performances used the rest of nature, props, stages, and so on. They were both touching and side-splitting. The collective effort was a triumph.

Being hugged by other facilitators and organizers is one of the most rewarding experiences for any facilitator, particularly as, only a few hours before, we had been in a crisis situation with few ideas for a solution. There were perhaps several factors which no doubt influenced the overall favourable outcome: geography (isolated, beautiful setting); a committed, active group of ecologically conscious individuals; involvement with other session leaders and opportunity to attend their sessions; a large group living together in a residence situation; and nature at our door (deer munched under our window each morning). However, we would like to believe that our success had to do with our 'sense of moment' or ability to evaluate need, our flexibility, and some 'on the spot' creativity.

Zen of Consumerism

The second activity which worked well, and continues to do so, is the *Zen of Consumerism: Waste "R" Us?* The primary purpose of the activity is to examine a wide variety of socio-environmental issues through the notion of 'place.' David Orr (1992, 126) has lamented that: "Often, the idea of 'place' has no particular standing in contemporary education...because we miss the immediate and mundane...and because place is a nebulous concept." We felt that learning and education should not be confined to the workshop venue or pedagogical institution but rather take place in the community at large. This was based on a recognition of the importance of location, and the utilizing the potential of 'place.' For place helps to stimulate interest and engages people through what it shows and the stories it tells.

During a workshop in Belleville, Ontario, we used two foreign-owned chain superstores, Toys "R" Us and K-Mart, as the learning 'place.' The purpose of this activity was primarily to examine the issue of waste and the politics of consumerism and their impact on the community. The participants were divided into pairs and given forms we had developed previously which asked them to identify fifteen items and allocate and/or describe them under the following headings: Need; Want; Amount/Type/Necessity of Packaging; Life Span; Where does It Go? As they entered the store, each pair chose a different set of aisles and began to allocate the various items under the proper columns. Upon returning to the original workshop venue, the group came together and the various pairs reported on their findings.

These went far beyond what we had expected. The groups not only discussed the immense amount of waste, but also the ruthlessness of advertising: in particular, the Disney Corporation, that 'double packages for advertising purposes.' They challenged 'green labelling' as being not so very green and raised the issue of what could be termed 'cultural imperialism.' Although the participants had all previously shopped in these stores and knew intellectually the negative impact they were having on the small businesses in the downtown core, actually 'seeing' and 'being' in place provided a new prism through which to view their community and initiated a more emotional reaction to the problem. Emotions, it can be argued, urge us into action more often than knowledge. Participants suggested that the next time we should include 'price and where things are made.' This would draw people into a broader discussion of globalization, fair wages and trade, human rights, and working conditions around the world.

In particular, all the participants had noticed the process of gender socialization. In Toys "R" Us there were aisles of pink kitchens, dollhouses, and beauty aids for the girls, and machine guns, trucks, and building materials for the boys. What we were being told is that girls are expected to nurture and construct themselves around a particular idea of beauty, while boys are expected to simultaneously construct the world for us and then destroy it. In this activity, we confirmed the potential of using 'place' as an environmental adult education tool. We realized together how much learning in place could stimulate ideas and transform people's perceptions of their immediate surroundings. At the closure of the activity one man remarked: "Now I not only have to wonder if this is good for my 'kid' but also have to wonder if it is good for the community and the world."

Diverse learning sites can refract the dominant representation of the world, providing people with a way to explain that representation in their everyday experiences. Using place can also highlight socio-political and gender inequities in ways in which staying in the classroom or workshop venue could never have done. Moreover, 'learning in place' can be about the global as well as the local for "nothing is so local that in one way or another, it is not a part of the 'global village' that this planet has become" (Viezzer 1992, 5).

The Wise Soil Activity

> There is no other way for humans to educate themselves...than through the instruction available through the natural world. (Berry 1988, 167)

The Wise Soil Activity did not begin with this name, but was dubbed it once we had reflected upon it and realized just how much the rest of nature had contributed to the discussions and other activities of this workshop. As in the activity outlined previously, learning in place was to be our salvation; nature as teacher and site of learning was to become more than just a 'nice idea.'

This workshop took place in a small town in northern Ontario, Rayside-Balfour. As always, we had been back and forth with the organizers, attempting to identify some key issues around which this workshop would revolve. The configuration of this group was somewhat different than what we were accustomed to as it included some young adults, secondary students in their mid-teens.

The workshop began very well but we soon began to notice that the young people were not speaking up in the group. As might be considered usual, the adults were doing all the talking. Although this had something to do with the newness of the topic, more importantly, it had to do with the imbalance of power between young people and adults. We decided to find out if the issues that the organizers had raised and felt were important were, in fact, those of the entire group. We put the young people into a separate group and the adults in other small groups and asked them to identify two or three pressing 'environmental issues.' As we had suspected, the adults came up with ideas such as soil erosion and transport while the young people came up with violence![7] The adults, ourselves included, were shocked by this. We had already thought about the issues of soil that had been raised by the organizers and had developed an idea for an outdoor activity, but how to relate this to violence against youth was another matter indeed.

> We wondered aloud whether we should break the group, which was getting into some difficult and necessary issues to go outside and get our hands into the soil? Trusting our instincts, which does not always work, we went off to the park near the lake. (Hall and Clover 1997, 744)

In mixed pairs of young and old, the participants were asked to collect samples of soil and potential soil makers. They brought their samples back and laid them carefully on display on a large piece of butcher paper we had spread on the ground. Each pair in turn knelt beside their collection of leaves, bark, grass, cigarette butts, worms, flowers, and so on, and told the group what they saw in their own samples and/or those of others.

The group, with no prompting, soon made links between the diversity of soil and the diversity of ideas and problems in the community. Issues such as "respect for difference, and patience with change, all issues which were critical to the community development process, were addressed 'naturally' in ways which we could not have predicted" (Hall and Clover 1997, 744). But, most remarkably, the group made the connection between soil erosion and violence against youth and this came in the form of urban rural tensions! Fertile soil was being transported from the community for use in city gardens. Older boys were coming from the urban areas, beating up on the local boys and harassing the girls.

By learning in, through, with, and about the rest of nature, we began to discover the community and its surrounding environment as a landscape of resistance, a source of regeneration, inspiration, beauty, and sensuality, a site of critique and organizing, and a site of envisioning and presenting networks of meaning. "Nature was truly both teacher and site of learning for all of us" (Hall and Clover 1997, 744).

These Failed Miserably

There were a few times when we failed, and failed miserably; when the activities we had created just fell flat. There are a number of reasons for this but the primary one was our mis-judgement of the vast knowledge of the participants on environmental issues and their complexity. Ironically, the idea of recognizing, respecting, and tapping into people's existing knowledge and experience is a fundamental principle of adult education;

but, in two particular instances we ended up, so to speak, with egg on our faces. Both times we were reminded to double-check our audience, their needs as well as backgrounds, before designing an activity. It is as easy to over-task a group as it is to force them to go where they have been many times, and the following is an example of our experience of the latter.

Analyzing Relations of Power: Global Environmental Justice

The purpose of this activity was to look at power and its relationship to poverty, race, gender, and the rest of nature, in a global perspective. We drew and put up on the wall, a map of the world, complete with continents. In the large group, participants identified the most pressing environmental issues facing the world today. These were written on brown boulders and placed around the continents. Then they broke into small groups, identified a *rapporteur*, and responded to the following questions: Who/What are most affected by environmental degradation? Why and how? Who/What are the least affected? What is the role of education?

All seemed to be going well; the map of the world looked impressive, everyone was eager to participate and contribute to the process, and because there were a few humourous responses to the first question, the activity had promise. However, within minutes after the small groups began discussing the questions, we, as facilitators became aware of discontent, and even rebellion, particularly in one of the groups. There were two reasons for this. First, the questions were too simplistic. As one woman pointed out, "everyone is affected by air pollution who lives in a city, so what is your point"? Second, these people were almost insulted by our questions, having repeatedly dealt with them in the past and in a more critical and 'adult' way. We soon realized that we had not prepared for our particular group of participants, many of whom had been a long time on-the-ground environmentalist and social activists. It seemed the only solution to save the day was by moving on to the next activity. We stopped the activity, admitted to the group that we had misjudged when preparing, and suggested that we move on. Everyone agreed.

This was an important learning experience for us, in that, creating an activity takes care and consideration. Never underestimate the knowledge of adults and never confuse 'childish' activities with fun and humour for they are not the same thing. Nevertheless, the final outcome of the workshop was very positive due to an activity immediately following in a nearby park (saved by the rest of nature again!). It was also 'rescued' by the wonderful stories participants told about doing environmental adult education and popular education. We will never forget our group in Halifax.

Web of Life

Another activity that did not work was the Web of Life. We had discovered variations of this activity or variations in several books on environmental education. On paper, it looked like an interesting and stimulating activity that would not require the energy needed to adapt it to suit a group of adults. The idea of the activity is to sit in a circle. Each person has a sign attached to her/himself with some element of the rest of nature written on it (i.e., sun, worm, tree, human). A ball of twine is thrown from one person to another, taking care to always keep hold of a piece of twine, until everyone has received the ball.

We soon began to notice that although people were having fun, laughing as the worm tossed the ball to the bird, they also seemed somewhat bored. When the last person had received the ball, we asked one of the participants to let go of the string and the web went slack and began to fall apart. We asked the painfully obvious question: What does this say about the inter-connectedness of life? There were some rather half-hearted responses, the usual "we are all connected," but things seemed to have gone flat. Finally, one woman looked up and said that if we had wanted to talk about inter-connections there was no need to engage in this childish game. She asked if we had ever taken a broom to a spider's web and watched the spider cling to the remnants of her work, her source of food, and possibly her life. Had we ever lost someone in our community and family and watched as people, numbed with grief, tried to pull the threads back together again? Had we ever seen the impact of drought on human existence, crops, and all other species—the inter-connection between life and climate? She talked of life as challenge, struggle, and a constant series of connections, inter-connections and disconnections. Using examples from human existence and the rest of nature, she made us realize we had underestimated our audience and trivialized an extremely remarkable issue. We sat humbled at the eloquence of the response, at the fundamental understanding of the different levels of inter-connectedness. The modified activity we use now reflects this wisdom and powerful critique.

Lessons Learned and Conclusions

With these lessons in hand, we go forward stumbling, colliding and, in the end, learning to dance without stepping on too many toes. (Keough 1996, 13)

Developing this resource and the activities included in it was first and foremost about developing the theory and practice of environmental adult education through a collective process with communities across Canada and worldwide. Through this process we learned a number of extremely valuable lessons, some more positive than others! We learned a number of important lessons:

1   It is very important to know your audience. Never underestimate people's knowledge: their ability to understand and describe/analyze the complexity of issues. Be prepared to be humbled and recognize the limitations of your own knowledge and ability.
2   Never underestimate the power of people's creativity—they will write a poem, they will put on a play, they will, they will!
3   Although activities can and should include elements of fun and humour, they must not be childish for this is not a 'game.' This work is about real people's lives and serious issues.
4   The activities one develops are contextual and, therefore, they should not be a series of prescribed techniques but rather, flexible, adaptable ways of working with people.
5   Linked to the above is the caution: Do not take the easy way out and simply use 'ready-made' activities.
6   Consciously include activities which are with, in, through, and about the rest of nature as an integral part of the programme—adults will go outside—and the rest of nature has its own abilities to teach, transform, stimulate, and engage.
7   There is an added value to creating a resource through a participatory process—there is more creativity, more ownership, more fun, and more challenges.
8   Theories and principles of any new form of education are emergent and changing; they are not constant, so add and subtract.

The people and challenges we met, the ideas we shared, and the excitement of success, especially the fun and joy of including the rest of nature into our framework of learning, have been our rewards for creating this resource book. Despite stepping on toes, we found this to be one of the best and most challenging learning experiences of our lives, and learned to dance. Through this collective, participatory process, we feel we have created a valuable environmental adult education and learning resource that contains the knowledge, concerns, creativity, and challenges of hundreds of people imprinted on each and every page.

Notes

1   There are actually two editions. The first is titled *The Nature of Transformation: Environmental Adult and Popular Education*, and was published in 1998. The second, published in 2000, was simplified to *The Nature of Transformation: Environmental Adult Education*.

2   This aspect became very apparent at the "Environment and Society: Education and Public Awareness for Sustainability" conference organized by UNESCO, in Thessaloniki, Greece, in 1997. Although there were literally hundreds of presentations and workshops, only two people, both of whom were connected to the Learning for Environmental Action Programme of the International Council for Adult Education, drew attention to the relevance of environmental adult education. 'Public' and 'society'—we reduced to 'school' and 'children.'

3   There are two more politically focussed texts using language of adult education which we found very useful: Ibikunle-Johnson and Rugumayo (1987) of Africa and Viezzer and Ovalles (1993) of Latin America.

4   This does not in any way imply that children do not have knowledge and experience. It simply means that with age and under normal circumstances, there is greater knowledge acquired through more experiences.

5   Over the years we used a number of different terms, such as adult environmental education, environmental popular education, environmental adult and popular education, and so on, before finally deciding upon the broader umbrella term 'environmental adult education.'

6   Our book also includes a section of activities which have been developed by the environmental adult educators from various parts of the world.

7   It is interesting to note that adults chose the more conventional ideas of what 'environmental issues' are, while the children seemed to view the environment more holistically and, therefore, chose the issues which most affected them in their environment.

References

Berry, Thomas. 1988. *The Dream of the Universe*. San Francisco: Sierra Club Books.
Bhasin, Kamla. 1992. "Alternative and Sustainable Development." *Convergence XXV 2*, 26–36.

Bishop, Anne. 1988. "Cartoons and Soap Operas: Popular Education in a Nova Scotia Fish Plant." *Convergence XXI,* 4, 27–32.

Bowers, C.A. 1991. "The Anthropocentric Foundation of Educational Liberalism: Some Critical Concerns." *Trumpeter 8,* 3, 102–107.

Burch, Mark. 1993. "Key Issues in Adult Environmental Education." In *Report of the Meeting of Experts.* Toronto: International Council for Adult Education.

Clover, Darlene. 1999. *Learning Patterns of Landscape and Life: Towards a Learning Framework of Environmental Adult Education.* Unpublished doctoral thesis. Toronto: University of Toronto, OISE.

Clover, Darlene, Follen, Shirley, and Hall, Budd L. 1998 and 2000. *The Nature of Transformation, Environmental Adult Education.* Toronto: University of Toronto, OISE.

Cohen, Michael. 1989. *Connecting with Nature: Creating Moments That Let Earth Teach.* Eugene, OR: World Peace University Press.

Hall, Budd and Clover, Darlene E. 1997. "The Future Begins Today. Nature as Teacher in Environmental Adult and Popular Education." *Futures 29,* 8, 737–747.

Howell, Calvin. 1989. "Emerging Trends in Environmental Education in the English-Speaking Caribbean." *Convergence XXII 4,* 45–53.

Ibikunle-Johnson, Victor and Rugumayo, Edward. 1987. *Environmental Education Through Adult Education.* Nairobi: African Association for Literacy and Adult Education.

ICAE/Ecologic. 1994. *Adult Environmental Education: A Workbook to Move from Words To Action.* Toronto: International Council for Adult Education.

Kane, Sean. 1998. *Wisdom of the Mythtellers.* Peterborough, ON: Broadview Press.

Keough, Noel. 1996. *Lessons from the Hinterland: Sustainability as if Reality Mattered.* Unpublished paper presented at the International Participatory Development Dialogue Symposium, Calgary, February.

Leal Filho, Walter and Villa Bandeira, Monica. 1995. Media and Environmental Education. *Convergence XXVIII,* 4, 55–60.

Nadeau, Denise. 1996. *Counting Our Victories, Popular Education and Organizing.* New Westminster, BC: Repeal the Deal Productions.

Orr, David. 1992. *Ecological Literacy: Education and the Transition to a Postmodern World.* Albany, NY: State University of New York Press.

Pike, Graham and Selby, David. 1988. *Global Teacher, Global Learner.* London: Hodder and Stoughton.

Thomas, Alan. 1991. Relationships and Political Science. In J. Peters and P. Jarvis (Eds.), *Adult Education, Evolution and Achievements in a Developing Field of Study.* Oxford: Peters, Jarvis and Associates.

Viezzer, Moema. 1992. "Learning for Environmental Action." *Convergence XXV,* 2, 3–8.

Wright, Tara. 2000. "Ecogames for Grown-Ups." *Alternatives 26,* 2, 41–42.

Contributors

Darlene Clover was co-organizer of the International Journey for Environmental Education and the Citizen's Treaty on Environmental Education for Global Responsibility and Sustainable Societies at the 1992 Earth Summit and International Coordinator of the Learning for Environmental Action Programme (LEAP) from 1995 to 2000. She is Assistant Professor in the Faculty of Education at the University of Victoria, British Columbia. Darlene's teaching and research areas include environmental adult education, women, and leadership, feminist pedagogy, the arts, and community activism.

Shirley Follen has an M.A. in adult education. She began her working career as community-college teacher of adult education, gender, and communications. She was the North American representative to the Learning for Environmental Action Programme (LEAP) from 1995 to 2000 and Coordinator of "Growing Jobs for Living," a community-based environmental adult education project.

José Roberto Guevara is an adult educator with a passion for process. He is most at home when he can work with communities and share the facilitation skills he developed while working with Philippine Educational Theatre Association (PETA), the Asian-South Pacific Bureau of Adult Education (ASPBAE), and the Centre for Environmental Concerns (CEC).

Budd Hall was Chair of the Department of Adult Education, Community Development and Counselling Psychology (AECDCP) at the Ontario Institute for Studies in Education, University of Toronto, for seven years. For twenty-five years Budd gained experience in grassroots adult education and participatory research through his work as Secretary-General of the Toronto-based International Council for Adult Education (ICAE). He is now Dean of the Faculty of Education at the University of Victoria, British Columbia.

Dip Kapoor was born in Calcutta, India, and became a permanent resident of Canada in 1993. He is currently Assistant Professor in the Department of Integrated Studies in Education at McGill University, Montreal. Dip was the president of HELP, an Edmonton-based NGO that works in partnership with tribal peoples in Orissa, India. Dip's research interests include international development theory, social change, social movements, and popular and nonformal adult and environmental adult education.

Karen Malone is currently UNESCO Asia-Pacific Director of the ☐rowing Up In Cities_ Project, and Associate Professor in the Faculty of Education, Language and Community Services, RMIT University, Australia. Her research and publication interests are social justice, equity, and participatory knowledge production and action. Karen has a strong philosophical orientation linked to principles of environmental sustainability and progressive social ecology, and her focus in education and environmental knowing encompasses both formal and nonformal training and workshops in capacity building for planners, government officials, community members, and children.

Rosa Muraguri-Mwololo is presently a development consultant in Nairobi who is completing a Ph.D. at Illinois University in the United States. She has vast direct development experience that spans nearly eighteen years, with the last ten years as a Senior Development Officer with the Canadian International Development Agency (CIDA) in Kenya. Rosa has facilitated numerous workshops in environmental adult education in rural Kenya, and has been a speaker at conferences around the world.

Joaquín Esteva Peralta has worked for over ten years in the area of rural development through environmental popular education. He is a trainer and researcher, working with Javier Reyes Ruiz, at the Centro de Estudios Sociales y Ecologicos (CESE, Centre for Social and Ecological Studies) in Pátzcuaro, Mexico, and has also widely published in the field of environmental popular education. Joaquín is a former Latin American regional representative to the Learning for Environmental Action Programme (LEAP).

Javier Reyes Ruiz has over fifteen years of experience in rural development, popular environmental education, and has published widely in this area. For over ten years, he has worked as a trainer and researcher at the Centro de Estudios Sociales y Ecologicos (CESE, Centre for Social and Ecological Studies in Pátzcuaro, Mexico. Javier is the Regional Coordinator of the environmental popular and adult education program of the Latin American Council for Adult Education (CEAAL).

Kerrie Strathy completed her MA in politics at the University of Guelph, and graduate studies in adult education at the University of Saskatchewan, Canada.

She has worked extensively in the area of international development, and received a CIDA (Canadian International Development Agency) scholarship to develop environment programs for women in Suva, Fiji. Kerrie managed a UNDP (United Nations Development Programme) environment project, became a founding member of WAINIMATE (Women's Association for Natural Medicinal Therapy), and its education advisor. Her interests are in the areas of participatory learning and action, community/non-formal environment education, gender and development, programme evaluation, and indigenous/traditional knowledge.

Salwa Tabedi worked with the Sudanese United Nations Scientific, Education and Cultural Organization in Khartoum for many years. She recently completed her doctoral thesis titled "The Impact of the Role of Urban Women in the Household on the Environment: A Case Study." She was LEAP (Learning for Environmental Action Programme) Regional Representative in the Arabic-speaking region from 1996 to 2000. Salwa is now living in the United States with her daughters.

Kesaia Tabunakawai worked in the forestry sector in Fiji after completing undergraduate studies in forestry, in India. She worked for Fiji Pine, and then became the information officer for Fiji's Forestry Department. Kesaia received a scholarship to do graduate studies in Forestry in Scotland and played a key role in WAINIMATE's (Women's Association for Natural Medicinal Therapy) development. In 1996 she joined the staff of the Worldwide Fund for Nature (WWF) South Pacific Programme Office in Suva. Her areas of expertise include biodiversity, conservation, and ethnobotany.

Sandra Tan holds her M.A. in adult education from the Ontario Institute for Studies in Education, University of Toronto (OISE/UT). Her academic interests include environmental racism, activism, and post-colonial discourse. Sandra recently gave birth to a boy.

Jan Woodhouse is currently Faculty Associate at Northern Illinois University, United States, and a doctoral student in the Adult Education Program. She considers this time as an interlude from nearly thirty years of work on the front lines of community-based organizational development and environmental education. Jan's focus of study is on international environmental adult education, especially as it involves women. Ancillary to this is a developing articulation of the theory and practice of place based pedagogy.

INDEX

Studies in the Postmodern Theory of Education

General Editors
Joe L. Kincheloe & Shirley R. Steinberg

Counterpoints publishes the most compelling and imaginative books being written in education today. Grounded on the theoretical advances in criticalism, feminism, and postmodernism in the last two decades of the twentieth century, Counterpoints engages the meaning of these innovations in various forms of educational expression. Committed to the proposition that theoretical literature should be accessible to a variety of audiences, the series insists that its authors avoid esoteric and jargonistic languages that transform educational scholarship into an elite discourse for the initiated. Scholarly work matters only to the degree it affects consciousness and practice at multiple sites. Counterpoints' editorial policy is based on these principles and the ability of scholars to break new ground, to open new conversations, to go where educators have never gone before.

For additional information about this series or for the submission of manuscripts, please contact:

Joe L. Kincheloe & Shirley R. Steinberg
c/o Peter Lang Publishing, Inc.
275 Seventh Avenue, 28th floor
New York, New York 10001

To order other books in this series, please contact our Customer Service Department:

(800) 770-LANG (within the U.S.)
(212) 647-7706 (outside the U.S.)
(212) 647-7707 FAX

Or browse online by series:
www.peterlangusa.com